Conceptual Physics
Laboratory Manual
by Paul Robinson
Illustrated by Paul G. Hewitt

 HarperCollins*CollegePublishers*

Conceptual Physics Laboratory Manual to accompany *CONCEPTUAL PHYSICS,*
Seventh Edition

Copyright © 1993 by Paul Robinson

ISBN 0-673-52252-0

· 93 94 95 5 4 3 2

"In the matter of physics, the first lessons should contain nothing but what is experimental and interesting to see. A pretty experiment is in itself often more valuable than 20 formulae extracted from our minds; it is particularly important that a young mind that has yet to find its way about in the world of phenonema should be spared from formulae altogether. In his physics they play exactly the same weird and fearful part as the figures of dates in Universal History."

— *Albert Einstein*

To the Instructor

This manual was designed to accompany Paul Hewitt's *Conceptual Physics*. This course may be your students' first encounter with physics. For the vast majority, it will be their last. Most enter with a limited laboratory experience in science.

Nevertheless, some of these students are the liberal arts students who will go on to become elementary school teachers — who will be asked to motivate young children in science. For this reason, this course could have a significant impact on not only their comprehension, but upon the outlook and attitude of children for years to come.

Typically, the college lab, which may be an elective, is a 2 or 3 hour block that meets once a week during the course of the semester. Presumably, the instructor will have 15 to 18 lab periods over the course of the semester. This allows about 2 or 3 lab sessions for each major unit covered in the text.

Conceptual Physics is *not* physics without math, but rather emphasizes *concepts* before *computation*. Similarly, the conceptual physics laboratory puts experiencing phenonema before trying to quantify them. For this reason, each unit has a variety of activities designed to both arouse student interest and acquaint them with the concepts. Some involve simple calculations. Although preferably performed by students, many can be done as demonstrations by the instructor. All "Activities" enable students to experience physics both as an activity and a process. The "Experiments" on the other hand, are more detailed investigations about a specific phenonemon and involve making quantitative measurements. Some involve the use of the computer.

This-manual format allows the instructor to design a variety of lab experiences to suit the student needs. Some instructors may choose to have their lab sessions consist entirely of activities; others may choose all experiments. Others will have students do two or three of the activities followed by a lab experiment.

Acknowledgements

Most of the ideas in this lab manual came from many of the physics instructors who share their ideas at *American Association of Physics Teachers* (AAPT) meetings that I have attended since my first year of teaching. This sharing of ideas and cooperative spirit is a hallmark of our profession.

Many more individuals have contributed their ideas and insights freely and openly than possibly can be mentioned here. I am grateful to all of them, and would like to thank especially Chuck Hunt, Scott Perry, Ann and John Hanks, Bill Papke, and Lowell Christensen of American River College who generously contributed so many of their ideas so freely; to Roy Unruh, University of Northern Iowa, Director of PRISMS, Sheila Cronin, Avon High School, CT, for her adaptions of CASTLE curriculum; Michael Zender, California State University, Fresno and Brian Holmes, California State University, San Jose, for their technical assistance and to Frank Crawford, University of California, Berkeley, and Verne Rockcastle, Cornell University, for their inspiration.

For reading and critiqueing my manuscript, I would like to thank John Hubisz, College of the Mainland, Texas, Al Bartlett, University of Colorado, David Ewing, Southwestern Georgia University, free lance physics editor Suzanne Lyons and Charlie Spiegel, California State College at Dominguez Hills. Thanks also to Dave Griffith and Paul Stokstad of PASCO Scientific for their professional assistance.

I am thankful to my AAPT colleagues Robert H. Good, California State University, Hayward, and Dave Vernier of Vernier Software for their suggestions on the best ways to integrate the computer in the physics laboratory. I am grateful to Skip Wagner for his creative work on the computer. I am especially indebted to my talented programmer and former student, Jay Obernolte, for developing software that accompanies this manual.

And thanks to my parents and children — David, Kristen, and Brian — for being so patient!

Most of all, I would like to express my gratitude to my mentor Paul Hewitt — not only for his illustrations and many suggestions, but for his optimism, and his support which have been a constant source of inspiration throughout my career.

Paul Robinson
Fresno
August, 1992

About the Author

Paul Robinson has been teaching physics since 1974 after getting his BA in Physics at the University of California, Santa Barbara. He completed his MA in Physics at Fresno State University in 1984. Despite his enthusiasm for teaching physics, he quickly realized his love for physics was not universal. While looking for new and innovative ways to lure students into his classroom, he met Paul Hewitt at a teachers' conference. Hewitt's fresh approach to physics was at the time considered controversial because it emphasizes thinking about the *ideas* of physics instead of doing computational physics problems.

Robinson invited Hewitt to be a guest lecturer in his classes. At the conclusion of each lecture, Robinson lay between two beds of nails as Hewitt crushed a large concrete block upon Robinson's chest with a sledge hammer. By the end of the day, people from all over town were coming to see this dramatic demonstration. Enrollments soared!

In 1983, Robinson was recruited to assist with the design and opening of Edison Computech — a magnet science, computer, and math school for grades 7–12 in Fresno. Encouraged by the administration to be innovative, he put a conceptual physics course at the *beginning* of the science sequence instead of at the *end*. This untraditional approach is now gaining nation-wide acceptance.

In 1987, he received the Presidential Award for Excellence in Science Teaching. He has also received the Distinguished Service Citation from the Northern California Section of the *American Association of Physics Teachers* and was the first high school teacher elected president of that association. He and his former student, Jay Obernolte, have developed computer software to accompany Robinson's lab manuals. In 1990, Robinson held the post of guest lecturer at American River College in Sacramento and he now teaches three levels of physics at Bullard High School in Fresno, California.

Table of Contents

Part 4: Sound

Part 5: Electricity and Magnetism

Part 6: Light

Part 7: Atomic and Nuclear Physics

Appendix Labs

Appendix

To the Student

We all know that to appreciate any game you need to know its rules. This is especially true of the physical world, where the rules are what physics is about. These rules are treated in your text. To further your understanding of physics, you need to know how to keep score. This involves observing, measuring, and expressing your findings numerically. That's what this manual is about.

Do physics and understand!

Name: _____ Section: _____ Date: _____

CONCEPTUAL **Physics**   **Experiment**

Perception

Is Seeing Believing?

Purpose
To experience a few optical illusions and to illustrate some of the limitations of the human senses as measuring devices.

Required Equipment and Supplies
meterstick
rotating disks as in Figures E and F or G
pencil or pen

Discussion

ANOTHER WARM DAY···
RELATIVELY SPEAKING

Can we trust our senses? Can we rely on the human senses of sight, hearing, smell, touch and taste to make accurate observations? Methods of measurement that rely entirely upon the senses are called *subjective* methods. Hot and cold, loud and soft are *subjective* terms. What seems cold to you may be quite comfortable to a polar bear. What is a comfortable volume on your stereo may be much too loud to your parents. Sometimes our senses fool us. Because early science relied heavily on the use of subjective methods scientific progress was slow.

The scientific approach is a way of answering questions about nature. During the 17th century, subjective methods were replaced by *objective* methods, using instruments to obtain measurements of greater precision. Objective methods minimize the effects of the observer on the results of an experiment. Of course, if we aren't careful, we can be fooled by our instruments too!

Is seeing believing? In this experiment you will perceive phenomena that demonstrate the need for objective methods.

Procedure

Step 1: Observe Figure A. Do the long slanted lines *appear* to be parallel? Are they?

Fig. A

_____

_____

Step 2: Observe the small black and white squares in Figure B. Do they appear to be the same size? Measure their sides and see.

Fig. B

_____

_____

Step 3: Look at the horizontal lines in Figure C. How do their lengths compare? Measure them and see.

Fig. C

_____

_____

Step 4: Look at the diagonal lines in Figure D. How do their lengths compare? Measure them and compare.

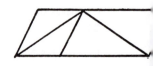

Fig. D

_____

Step 5: Adjust the speed of an electric motor to spin a disk like Figure E at various speeds about 4 to 15 revolutions per second. If an electric motor is not available, try to spin the disk on your pencil. Do you see colors? If so, which ones? Record your observations.

_____

_____

Step 6: Rotate the disk illustrated in Figure F on your pencil about 2 or 3 revolutions per second. What do you observe?

_____

_____

Step 7: Stare at the center of the disk illustrated in Figure G for 30 seconds as you slowly spin the disk at about 1-2 revolutions per second. Then stare at the palm of your hand. Try again but stare at your surroundings after staring at the disk. What do you observe?

_____

_____

Step 8: Study the people in Figure H. Which one appears tallest? Measure the people in the figure and see.

_____

_____

Fig. H

Fig. E

Is Seeing Believing?

3

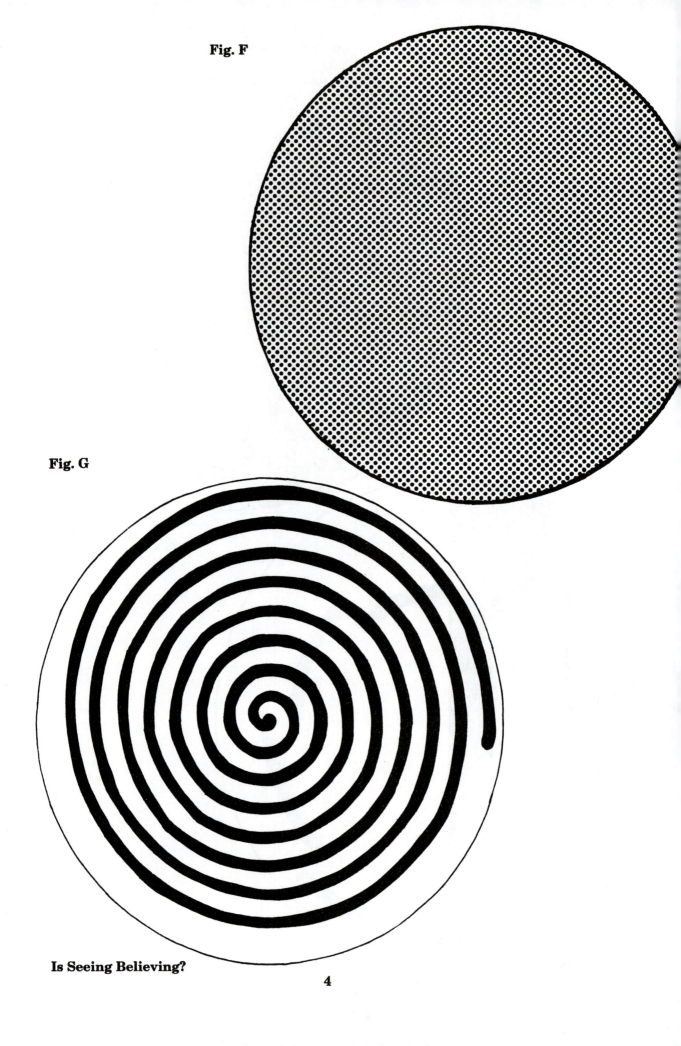

Fig. F

Fig. G

Is Seeing Believing?

4

Step 9: Hold your hands outstretched, one twice as far from your eyes as the other, and make a casual judgement as to which hand looks bigger. Most people see them to be about the same size, while many see the nearer hand to be slightly bigger. How about you?

_____

_____

After you have done this and made your judgement, overlap your hands slightly and view them with one eye closed. How do they appear now?

_____

_____

Summing Up

What does your experience with the perceptual activities in this laboratory tell you about the reliability of human senses as measurement tools?

_____

_____

_____

_____

_____

_____

Name: _____ Section: _____ Date: _____

Amassing a Penny's Worth

Purpose
To show how precision measurements can reveal information that might otherwise go unnoticed, and how to make a histogram.

Required Equipment and Supplies
beam balance
20 pennies: 10 dated before 1982; 10 dated after 1982
graph paper

Discussion
Making measurement sometimes leads to important discoveries that might otherwise go unnoticed. In this activity you will measure the mass of some pennies and plot a histogram of your results. The results may be surprising.

The *mass* of an object refers to the amount of matter it possesses. The *weight* of an object, on the other hand, refers to the force of gravitation on the object. Even though the mass of something remains constant, its weight can vary, for gravitational force depends on distance from the earth's center. We weigh a little bit less at the top because we're farther from the center of the earth. So weight depends on location whereas mass doesn't. Mass is a property only of the object.

It is customary to measure the weight of things by how much they stretch a spring. A *spring scale* measures weight. It is customary to measure the mass of an object by comparing it to a *standard mass*. Masses may be compared with a *beam balance*. Whereas a spring scale will give different readings in regions of different gravitation, a beam balance will accurately compare masses anywhere. A kilogram will balance a kilogram just as accurately at the top of a mountain as it will at sea level. Location makes no difference with a beam balance.

Procedure
Step 1: Adjust your beam balance so that both pans balance with zero load. This is *zeroing* the balance. After carefully zeroing the balance, find the mass of a penny to the nearest hundredth of a gram and write down your measurement. Repeat for all of the pennies.

Step 2: Note that the pennies most likely do not have identical masses. To see if there is any pattern to the data, make a *histogram*. A histogram is a plot of the frequency of each measurement in a series of measurements. Label the horizontal axis as the mass and the vertical axis as the number of times that measurement occurred (sometimes called the "frequency" axis). The horizontal axis should range from the smallest to the largest mass in tenths of a gram. Number the vertical axis from 0 to 10. Place a dot above the value on the horizontal axis for each mass measurement.

PENNY	MASS(g)
1	
2	
3	
4	
5	
6	
7	
8	
9	
10	
11	
12	
13	
14	
15	
16	
17	
18	
19	
20	

Amassing a Penny's Worth

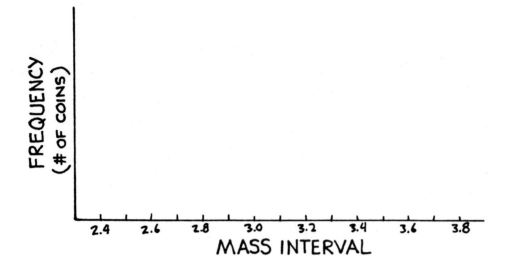

Summing Up

1. Does your histogram reveal any information about the pennies that was not apparent by just looking at them? Do the coins *look* the same?

_____

_____

2. How do you account for the measured differences of the masses of the coins?

_____

_____

3. Can you think of other items that might exhibit the same characteristics?

_____

_____

CONCEPTUAL **Physics**

Accelerated Motion

Tin Pan Alley

Purpose
To investigate the distance of free fall during equal time intervals.

Required Equipment and Supplies
six 1/2-inch hex nuts
kite string
masking tape
meterstick
pie pan

Discussion
Consider a long piece of string with hexagonal nuts taped at regularly-spaced intervals — say 10 centimeters apart. With the help of a ladder, suppose you hold one end of the string near the ceiling, with the rest of the string hanging vertically over a pie pan on the floor below. If you drop the string, the nuts will make a clanging sound as they hit the pan below. Will they hit at evenly-spaced time intervals? A little thought reveals that because the nuts accelerate as they fall, the time between clangs becomes less and less as the nuts strike the pan.

WILL EQUAL DISTANCES OR UNEQUAL DISTANCES YIELD EQUAL TIME INTERVALS WHEN THE NUTS HIT THE PAN ?

The goal of this activity is to tape hexagonal nuts onto a piece of string in such a way that when the string is held vertically and dropped, the nuts will hit the pie pan at equal *time* intervals. This experiment does not require the use of standard units of time, such as seconds or minutes. We will simplify calculations by using the elapsed time between nuts hitting a pan. We call this unit a "beat" because the sense of rhythm is used to judge whether or not the nuts hit the pan in equally spaced time intervals.

Procedure
Step 1: Use a string as long as the highest available location from which the string can be dropped.

Step 2: Tie the first nut to one end of the string. Calculate the appropriate positions for at least 5 other nuts so they strike the pan in equal intervals.

Step 3: Invert the pie pan on the floor so the nuts will make a clang upon impact. Hold the string so the first nut just rests on the pan and the second is at least 10 cm above the pan. Let go and listen to the "clangs" carefully! Do the clangs occur at equal time intervals?

_____

_____

BEAT (NUTS)	BEAT^2	TOTAL DISTANCE
1	1	1d
2	4	4d
3	9	9d
4	16	16d
5	25	25d

Hint 1: If the falling nuts are *equally spaced* along the string, the nuts speed up (accelerate) as they fall due to gravitation. Thus, as time goes on, the nuts take shorter times to fall a given distance. The time between clangs becomes less as the nuts strike the pan. To remedy this, shouldn't the distance between nuts *increase* with increasing height to compensate for increases in speed as they fall?

_____

_____

Hint 2: The distance an object falls in time t is $\frac{1}{2}gt^2$. Since $\frac{1}{2}g$ is constant, $d \sim t^2$.

The distance is equal to a constant $\frac{1}{2}g$ multiplied by t^2. The constant is the first distance between the first two nuts. The time it takes the first nut above the pan (the second nut) to hit the pan is one *beat*.

Try it and see. How do the clangs sound now?

_____

_____

Summing Up

Describe your spacing pattern that results in the clangs occuring at equal time intervals.

_____

_____

_____

_____

_____

_____

Tin-Pan Alley

CONCEPTUAL **Physics** ————————————————————— | **Activity** |

Free Fall and Weightlessness

Styrofoam Astronauts

Purpose
To observe the effects of gravity on objects in free fall.

Required Equipment and Supplies
2 Styrofoam or paper cups
2 long rubber bands
2 washers or other small masses
masking tape
large paper clip
water

Discussion
It is commonly believed that *since* astronauts aboard an orbiting space vehicle *appear* to be weightless, the pull of gravity upon them is *zero*. This condition is commonly referred to as "zero — g". While it is true that they *feel* weightless, gravity *is* acting upon them (Chapter 8 in *Conceptual Physics*). Gravity at space shuttle heights is in fact only about 10% less than at the earth's surface.

The key to understanding this condition is realizing that both the astronauts and the space vehicle are in free fall. It is very similar to how you would feel inside an elevator with a snapped cable! The primary difference between the runaway elevator and the space vehicle is that the runaway elevator has no *horizontal velocity* (relative to the earth's surface) as it falls toward the earth, so it eventually hits the earth. The horizontal velocity of the space vehicle ensures that as it falls *toward* the earth it also moves *around* the earth. The combination of motions (tangential and downward) results in it falling without getting closer to the earth's surface. Both the runaway elevator and the orbiting space vehicle are in free fall.

Procedure
Step 1: Elevators, especially those in tall buildings, are capable of changing the weight you feel or your *apparent* weight. When an elevator first starts to move up or down, your body senses the change in speed. Shortly afterwards, it moves at more or less constant speed until it begins to slow down to a stop. Weigh yourself on a bathroom scale in a motionless elevator. Observe what happens to the reading on the scale (your apparent weight) as the elevator:

a) accelerates upward.

_____

b) accelerates downward.

_____

c) moves upward at a constant speed.

_____

d) moves downward at a constant speed.

1. What would the scale reading be if the elevator cable(s) broke?

Step 2: Knot together two rubber bands to make one long rubber band. **Fig. A**
Knot each end around a small steel washer, and tape the washers to the
ends. Poke or bore a small hole about the diameter of a pencil through the
bottom of a Styrofoam or paper cup. Fit the rubber bands through the
hole from the inside. Use a paper clip to hold the rubber bands in place
under the bottom of the cup (see Figure A). Hang the washers over the lip
of the cup. The rubber bands should be under enough tension to keep the
bands taut but not so much as to flip them into the cup.

Step 3: Drop the cup from a height of about 2 m.

2. What happens to the washers?

Fig. B

Step 4: Remove the rubber bands from the cup and fill the cup half-full
with water, using your finger as a stopper over the hole. Hold the cup
directly over a sink or waste basket. Drop the cup into a sink or waste
basket. What happens to the water as the cup falls?

Step 5: Repeat Step 3 for a second cup half-filled with water with two
holes poked through its sides (Figure B).

Summing Up
3. Explain why the washers acted as they did in Step 2.

4. Explain why the draining water acted as it did in Steps 3 and 4.

CONCEPTUAL **Physics** | **Activity**

Free Fall and Friction

What a Drag!

Purpose
To compare the effects of air friction on a falling body.

Required Equipment and Supplies
sheet of paper
coin
book

Discussion
You know it takes longer for a leaf to fall to the ground than a rock. Is it because the rock is heavier than the leaf? Is it due to the air? Try this activity and see for yourself!

Procedure
Drop a sheet of paper from one hand, and a coin from the other. Release them at the same time. Which reaches the ground first?

_____

Why?

_____

_____

Now crumple the paper into a small, tight wad and drop it and the coin again together. Explain the difference observed. Will they fall together if dropped from a second-, third-, or fourth-story window? Try it and explain your observations.

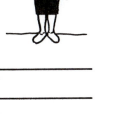

_____

_____

_____

Drop a book and a sheet of paper and notice that the book has a greater acceleration — g. Place the paper beneath the book and it is forced against the book as both fall, so both fall with the acceleration g. How do the accelerations compare if you place the paper on *top* of the raised book and then drop both? You may be surprised, so try it and see. Then explain your observations.

_____

_____

_____

CONCEPTUAL **Physics**

Action-Reaction

Tug-of-War

Purpose
To introduce the concept of action-reaction force pairs.

Required Equipment and Supplies
two large demonstration scales (about 50 or 100-N capacity)
string
two table clamp pulleys
two 1-kg masses and hangers
one 2-kg scale

Discussion
Think of a push or a pull on something — a bat on a ball, hands pulling on a rope, a foot on a ball. Notice that any situation you think of involves a *pair* of interacting objects. Forces always occur in pairs, often called "action-reaction" pairs. Chapter 4 in *Conceptual Physics* concludes with the observation that you cannot touch without being touched — Newton's third law.

Procedure
Pull gently on your lab partner's hand. Have him or her resist your pull. If you pull with a force of 10 N, how much force does your partner exert?

_____

Suppose your pull on your partner is called the action force. Identify the reaction force.

_____

Now repeat using a spring scale — then with two spring scales. How much is the action force? The reaction force?

action force: _____   reaction force: _____

Hook your scale to the wall. Pull with a force of 20 N. Identify the action-reaction pair of forces.

_____

Now set up a spring scale with two equal weights hanging over pulleys as shown in Figure A. What is the reading of the scale — as evidenced by a pair of forces that stretch the spring inside?

_____

What is the net force on the scale itself — as evidenced by its state of motion?

_____

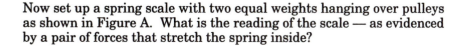
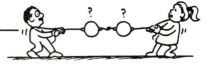

Tug-of-War

CONCEPTUAL **Physics** ———————————————————— | **Activity** |

Energy and Power

Powerhouse

Purpose
To determine the power that can be produced by various muscles of the human body.

Required Equipment and Supplies
bleachers and/or stairs
stopwatch
meterstick
weights
rope

Discussion
Power is usually associated with mechanical engines or electrical motors (Chapter 6 in *Conceptual Physics*). Many other devices also consume power to make light or heat. A lighted incandescent bulb may dissipate 100 watts of power. The human body dissipates about 100 watts as it converts the energy of food to heat. The human body is subject to the same laws of physics that govern mechanical and electrical devices.

The different muscle groups of the body are capable of producing forces that can act through distances. Work depends upon the force and the distance when both have the same direction. A person running up stairs does work against gravity. The work done can be measured by the person's weight multiplied by the vertical distance moved (not the distance along the stairs). This work per time it takes is power. When work is in newton–meters (or *joules*), and time is in seconds, power is in *watts*.

Procedure
Step 1: Select an activity from the following list:

- Lift a mass with your wrist, forearm, arm, foot, or leg only.

- Do pushups, sit-ups, or some other exercise.

- Run up stairs or bleachers.

- Lift a weight with a rope.

- Jump with or without weights attached to your body.

Step 2: Perform the activity, and record in Table A. Express force in newtons, distance moved by the force in meters, and time required in seconds. Calculate the power in watts. Show your calculations.

Summing Up

1. Suppose you lifted a 25-N brick vertically one meter using your arm. How much work do you do *lifting* the brick? How much work do you do on the brick *lowering* it one meter?

_____

_____

2. Which of the activities done in class produced the greatest *power*? Which muscle groups were used in this activity?

_____

3. Did the activity that used the largest force result in the largest power produced? Explain how exerting a large force can result in small power.

_____

4. Can a pulley, winch, or lever increase the *rate* at which a person does work? Be careful of the wording of this question, and explain your answer.

_____

_____

Table A

DATA TABLE   Power = $\dfrac{F \cdot d}{t}$

FORCE					
DISTANCE					
TIME					
POWER					

CONCEPTUAL **Physics** | **Activity**

Center of Gravity and Stability

Point of No Return

Purpose
To experience how the CG is above an area of support prevents toppling.

Required Equipment and Supplies
meterstick
string
plumb bob or suitable weight
stiff paper
tape
scissors
tennis racquet, baseball bat, other asymmetrical objects

Discussion
In order to maintain your balance, you must keep your CG above a point of support or you will fall over or topple. The Leaning Tower of Pisa does not topple because its CG is above points of support.

Stand with your heels and back against a wall and try to bend over and touch your toes. You'll find you have to stand away from the wall to do so without toppling over (Figure 7.23 in *Conceptual Physics*).

Procedure
Step 1: Compare the minimum distance of your heels from the wall when you didn't topple with those of a friends of the opposite sex.

1. Who can generally touch their toes with their heels nearer to the wall — men or women?

The center of gravity of a uniform object such as a meterstick is at its midpoint, for the stick acts as though its entire weight were concentrated there. When that single point is supported the whole stick is supported and has no tendency to rotate or turn. Balancing an object provides a simple method of locating its center of gravity. Find the center of gravity of a baseball bat by balancing it.

2. Where would you connect a string to the bat so when the bat is suspended it hangs horizontally?

Step 2: The center of gravity of a freely suspended object lies directly beneath the point of suspension. If a vertical line is drawn through the point of suspension, the center of gravity lies somewhere along that line. To determine exactly where it lies along the line, we only have to suspend the object from some other point and draw a second vertical line through that point of suspension. The center of gravity is located where the two lines intersect.

Point of No Return

Using a string and a plumb bob or some other handy weight, find the center of gravity of a tennis racquet (or some other irregular object) by suspending it from several different points. You may want to tape some stiff paper on one side of the racket so you can sketch your vertical lines on it.

3. Does the racket balance when suspended from this point? Explain.

_____

_____

Summing Up

4. On the average and in proportion to height, which sex has the lower center of gravity? Why?

_____

_____

_____

_____

CONCEPTUAL **Physics** | **Activity**

Torque

Torque Feeler

Purpose
To illustrate the qualitative differences between torque and force.

Required Equipment and Supplies
meterstick
meterstick clamp
one 1-kg mass
mass hanger

Discussion
Torque and force are sometimes confused because of their similarities. Their differences should become evident in this activity.

Procedure
Step 1: Hold a meterstick between your thumb and index finger at the 5 cm mark. With the stick held horizontally, position the mass hanger at the 10-cm mark, and suspend the 1-kg mass from it. Rotate the stick to raise and lower the free end of the stick. Note how hard or easy it is to raise and lower the free end of the stick.

Step 2: Move the mass-hanger to the 20-cm mark. Rotate the stick up and down about the pivot point (your index finger) as before. Repeat this procedure with the mass at the 40-cm, 60-cm, 80-cm, and 95-cm marks.

Summing Up
1. Does the stick get easier or harder to rotate as the mass gets farther from the pivot point?

_____

2. Does the weight of the mass *increase* as you move it away from the pivot point (your index finger)?

_____

3. Assuming the weight of the mass doesn't change, why does the difficulty increase in rotating the stick when the mass is positioned farther from the pivot?

_____

_____

CONCEPTUAL **Physics** | **Activity** |

CG and Torques

Hanging Out

Purpose
To learn that the CG of a static system is above a point of support.

Required Equipment and Supplies
Styrofoam or wood blocks or metersticks
baseball bat
tennis racquet

Discussion
Why doesn't the Leaning Tower of Pisa topple? Objects topple when their center of gravity is beyond the area of support (Figure 7.20 in your text). How can you stack four identical blocks with a maximum overhang without toppling over?

Procedure
Stack four Styrofoam blocks, about 5 x 20 x 40-cm in size on top of each other on the lab table (if unavailable, use four books or four metersticks). Try starting with one block, then two, three, and so on. Using only one block at each elevation, what arrangement of blocks results in the maximum possible overhang? Make a sketch of your final arrangement. Label the distance each block extends beyond the block below it.

Summing Up
Can you propose a theory which predicts the maximum overhang?

_____

_____

_____

_____

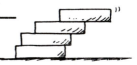

CONCEPTUAL **Physics** | **Activity**
Rotational Inertia

Rotational Derby

Purpose
To observe how objects of different shapes and mass roll down an incline and how rotational inertia affects rotation rate.

Required Equipment and Supplies
smooth, flat 2-m long board
ring stand or metal support with clamp and rod
balance
meterstick
3 solid steel balls of different diameter (3/4" minimum)
3 empty cans of different diameter, with both top and bottom removed
3 unopened cans of different diameter, filled with non-sloshing contents (such as chili or ravioli)
2 unopened soup cans filled with different kinds of soup, one liquid (sloshing—such as *Beef Broth*) and one solid (non-sloshing—such as *Bean with Bacon*)

Discussion
Why is most of the mass of a flywheel or a gyroscope or a Frisbee concentrated at its outer edge? Does this mass configuration give these objects a greater tendency to resist changes in rotation? How does this "rotational inertia" differ from the inertia studied when we investigated linear motion? Keep these questions in mind as you do this experiment.

The *rotational speed* of a rotating object is a measure of how fast the rotation is. It is the size of the angle turned per unit of time and may be measured in degrees per second or in radians per second. (A radian is a unit similar to a degree, only bigger; it is slightly larger than 57 degrees.) The *rotational acceleration*, on the other hand, is a measure of how quickly the rotational speed changes. It is measured in degrees per second each second or in radians per second each second. The rotational speed and the rotational acceleration are to rotational motion as speed and acceleration are to linear motion.

You are going to "race" a variety of solid steel balls, cylinders, and cans of food down a ramp. Even though these objects *fall* to the ground with the same acceleration when dropped, they do not necessarily *roll* down an incline at the same acceleration. This is because they may have different rotational inertias (see Figure 7.3 in the text).

Procedure
Step 1: Use a flat, smooth board about 2 meters long as a ramp. Place the board at an angle of about 10 degrees. You may use a ruler to release the objects quickly and cleanly.

1. Gravity produced an unbalanced torque that caused the objects to roll faster and faster — that is, to *rotationally accelerate*. Which objects had the greater *rotational inertia* — that is, the greater resistance to rotational acceleration?

2. How did the relative times for the objects to roll from the top of the incline to the bottom compare to the rotational inertias of the objects? (Formulas for the rotational inertias of various objects are shown in Figure 7.3 in the text.)

_____

_____

_____

_____

3. Did the size and mass of similar types of objects determine the winner?

_____

_____

Step 2: Consider a race between an object that slides down a friction-free incline and an object that rolls down the same incline with friction (without friction, an object would slide without rolling!) With no rotational inertia to contend with, the freely sliding object will win. Things sliding without friction will beat rolling things when the slope and distances are the same. Now try racing two soup cans, one with liquid sloshing contents and the other with solid non-sloshing contents. Predict the winner. Try it and see if your predictions are right.

predicted winner: _____

actual winner: _____

Summing Up
Can you explain the results?

_____

_____

_____

_____

_____

_____

Rotational Derby

CONCEPTUAL **Physics** ━━━━━━━━━━━━━━━━━━━━━━━━━━━━━━━━ | **Activity** |

Levers and Torques

Name that Lever

Purpose
To identify levers on a ten-speed bicycle.

Required Equipment and Supplies
ten-speed bicycle
ruler

Discussion
Many mechanical devices use levers. Levers can increase force by decreasing distance moved, or decrease force and increase distance. Although the work done by a device can never be more than the work or energy invested in it, levers make work *easier* to accomplish for a variety of tasks. A relatively small downward force exerted on the longer side of a lever can produce a relatively large upward force on the shorter side. The force exerted, F_e, and the distance through which this force moves, d_e, will equal the resistance force F_r and the distance it is moved, d_r. That is,

$$\frac{F_e}{d_e} = \frac{F_r}{d_r}$$

The mechanical advantage of a lever is defined as the ratio of effort distance to resistance distance or the resistance force to the effort force.

$$MA = \frac{d_e}{d_r} = \frac{F_r}{F_e}$$

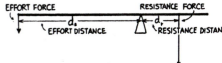

When a small input force moves through a relatively large distance and the larger output force moves through a relatively small distance, the mechanical advantage, MA, is greater than 1. When a relatively large input force moves through a small distance and the smaller output force moves through a larger distance, the MA is less than 1. In this case there is a gain in distance (or speed) instead of force.

Fig. A

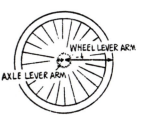

The locations of lever arms and fulcrums are not always as obvious as they are for a see-saw. In the see-saw the effort and resistance forces are parallel. Figure A depicts a claw hammer pulling a nail. In this example the forces are not parallel. The fulcrum is the point about which the hammer pivots or rotates as the nail is pulled. One lever arm is the perpendicular distance from the nail to the fulcrum. The other is the distance from the hand to the fulcrum.

A wheel and axle is a type of lever as shown in Figure B. In this case, the fulcrum is the point about which the wheel rotates, at the center of the axle. The lever arms are the radii of the wheel and the axle.

Fig. B

The tuning knobs of a microscope are levers. Knobs work very much the same way as a wheel and axle. Have you ever noticed that the coarse tuning knob is harder to turn than the fine tuning knob? The purpose of a coarse tuning knob is to move the microscope objective a large distance quickly, whereas the purpose of the fine tuning knob is to move the microscope objective a small distance slowly. The coarse tuning knob has a smaller radius than the fine tuning knob while both their axles have the same radius. For the coarse tuning knob, a larger effort force is necessary to produce a larger displacement of the microscope objective. A smaller force is needed on the fine tuning knob resulting in a smaller displacement.

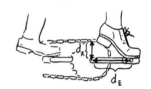

You are going to make sketches of levers and their corresponding lever diagrams. The lever diagram is more abstract, having a straight line with a triangular fulcrum, with arrows in the direction of the forces applied *on the lever*. Be sure to distinguish between the effort and the resistance forces. Draw dashed lines along the direction of the effort or applied forces and solid lines along the direction of the resistance forces.

Procedure

Step 1: Carefully study a ten-speed bicycle. On the drawing in Figure C, identify at least three different levers. Draw the corresponding lever diagrams to the right of your sketch of the lever. Label the fulcrum, the effort forces, the resistance forces, and their respective lever arms below.

Fig. C

Name that Lever

Step 2: Study the sketches and their corresponding lever diagrams to identify whether each lever is designed for a gain in *force* or for a gain in *distance* or *speed*. Note whether the MA provides a gain of distance or speed.

Summing Up

1. Is the chain of a bicycle a type of lever? Likewise, are the brake and gear-shift cables types of levers?

_____

2. Are the bicycle pedals a lever?

_____

_____

3. In what ways can a wheel and axle be considered a lever?

_____

_____

Name that Lever

CONCEPTUAL Physics

Satellite Motion

Being Eccentric

Purpose

To get a feeling for the shapes of ellipses and the locations of their foci.

Required Equipment and Supplies

20 cm of string
2 thumbtacks or small pieces of masking tape
pencil
graph paper

Discussion

The path of a planet around the sun is an ellipse. The path of Halley's comet around the sun is also an ellipse. An ellipse is an oval-like curve and is defined as the locus of all points the sum of whose distances from two fixed points (called *foci*) is a constant. One way to draw an ellipse as Figure 9.64 in your test suggests is to place a loop of string around two push pins and pull the string taut with a pencil. Then slide the pencil along the string, keeping it taut. To keep the string from getting twisted, it helps if you make the top and bottom half in two separate operations.

Procedure

Step 1: Using graph paper, a loop of string, and two push pins (or a small piece of masking tape instead), draw an ellipse. Label each focus of your ellipse (the location of the push pins).

Step 2: Repeat twice, using a different focus separation distance. The greater the distance between foci, the more eccentric is the ellipse.

Step 3: A circle is a special case of an ellipse. Determine the positions of the foci that give you a circle, and construct a circle.

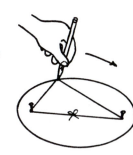

Step 4: Another special case of an ellipse is a straight line. Determine the positions of the foci that give you a straight line.

Summing Up

1. Which of your drawings is closest to the earth's orbit around the sun?

_____

2. Which of your drawings is closest to the orbit of Halley's comet around the sun?

_____

Being Eccentric

CONCEPTUAL **Physics**

Experiment

Accelerated Motion

Split Second

Purpose
To measure reaction time and the role it plays in a variety of situations.

Required Equipment and Supplies
dollar bill
centimeter ruler
LabTools software or equivalent
computer

Discussion
Reaction time is the time interval between receiving a signal and acting on it — for example, the time between a tap on the knee and the resulting jerk of the leg. Reaction time often affects the making of measurements, as, for example, using a stopwatch to measure the time for a 100-m dash. The watch is started after the gun sounds and is stopped after the tape is broken. Both actions involve the reaction time.

Procedure
Step 1: Hold a dollar bill so that the mid-point hangs between your partner's fingers. Challenge your partner to catch it by snapping his or her fingers shut when you release it.

The distance the bill will fall is found using

$$d = \tfrac{1}{2}at^2$$

Simple rearrangement gives the time of fall in seconds

$$t^2 = \frac{2d}{g}$$

$$t = \sqrt{\tfrac{2}{980}}\sqrt{d}$$

$$t = 0.045\sqrt{d}$$

(for d in cm, t in s, we use $g = 980$ cm/s^2)

Step 2: Have your partner similarly drop a centimeter ruler between your fingers. Catch it and note the number of centimeters that passed during your reaction time. Then calculate your reaction time using the formula

$$t = 0.045\sqrt{d}$$

where d is the distance in centimeters.

reaction time = _____

Split Second

Optional Computer Activity

Step 3: In this activity you will measure your reaction time using a different method. If Available, your instructor will install an interface box on the computer for you. Select "Reaction Timer" from the "Chronometer" program on the disk *LabTools*. Measure your reaction time several times with your fingers on button "0" of the interface box. Try responding to the audio stimulus as well as the visual one (or both). Then repeat measuring the time it takes you to depress the switch button starting with your hand flat on the table. Record your data and compute the average value of your reaction times.

INTERFACE BOX

Summing Up

1. Do you think your reaction time is always the same? Is your reaction time different for different stimuli?

2. Suggest possible explanations why reaction times are different for different people.

3. Do you think reaction time significantly affects measurements you might make using instruments for this course? How could you minimize its role?

4. What role does reaction time play in applying the brakes to your car in an emergency situation? Estimate the distance a car travels at 100 km/h during your reaction time in braking.

5. Give examples in which reaction time is important in sports.

6. Did your average reaction times differ when you held your fingers directly over the button compared to when they were positioned on the table? How is this comparable to driving a car?

CONCEPTUAL **Physics** ━━━━━━━━━━━━━━━━━━━━━━━━━━━━━ **Experiment**
Graphical Analysis of Motion

Blind as a Bat

Purpose
To make qualitative interpretations of motions from graphs.

Required Equipment and Supplies
sonic ranger
Super Sonic Plus software or equivalent
computer
masking tape
marking pen or pencil
ring stand
large steel ball
heavy beach ball
dynamics cart
can of soup
board
pendulum clamp
printer (optional)

Discussion
A bat can fly around in the dark without bumping into things by sensing the echoes of squeaks it emits. These squeaks reflect off walls and objects, return to the bat's head, and are processed in its brain to provide the location of nearby objects. The automatic focus on some cameras works on very much the same principle. The sonic ranger is a device that measures the time that ultra high frequency sound waves take to go to and return from a target object. The data are fed to a computer where they are graphically displayed. The program can display the data in three ways: distance vs. time, velocity vs. time, and acceleration vs. time. Imagine how Galileo would marvel at such technology!

Procedure
Step 1: Your instructor will install the sonic ranger on the computer for you. Check to see that it appears to be hooked up and operating properly. Place the sonic ranger on a desk or table so that its beam is about chest high. (Note: Sometimes sonic rangers do not operate reliably on top of computer monitors.)

Select the option that plots distance vs. time. Set the graph duration to "Continuous" and the maximum range to 4 meters. The sonic ranger has a *minimum* range of about 0.4 meters. Readings within this range will be erratic. Make small pencil marks or affix a piece of string to the floor in a straight line from the sonic ranger. Point the sonic ranger at a student standing at the 5 meter mark. Mark where the computer registers 0 m, 1 m, 2 m, etc. Set the maximum range appropriately between 4 to 6 meters.

Part A: Analyzing Motion Plots

Step 2: Stand on the 1-meter mark. Face the sonic ranger and watch the monitor. Walk away from the sonic ranger slowly and observe the real-time plot. Repeat, walking away from the sonic ranger faster, and observe the graph.

1. Make a sketch (or printout if a printer is available) of each graph. How do the graphs compare?

_____

Step 3: Stand at the far end of the string. Slowly approach the sonic ranger and observe the graph plotted. Repeat, walking faster, and observe the graph plotted.

2. Make a sketch of each graph. How do the graphs compare?

_____

Step 4: Walk away from the sonic ranger slowly; stop; then approach the sonic ranger quickly.

3. Sketch the shape of the resulting graph. How do the slopes change?

_____

Step 5: Repeat Step 2, but select the option to display to plot velocity vs. time.

4. Make a sketch of the graph. How do the distance vs. time and velocity vs. time graphs compare?

_____

Blind as a Bat

Step 6: Repeat Step 3, but select the option to display to plot velocity vs. time.

   5. Make a sketch of the graph. How do the two new graphs compare?

_____

Step 7: Repeat Step 4, but select the option to display to plot velocity vs. time.

   6. Sketch the shape of the resulting graph. How do the two new graphs compare?

_____

Part B: Analyze Motion on an Incline

Step 8: Set up the sonic ranger as shown in Figure A, to analyze the motion of a soup can, dynamics cart or a large steel ball that is rolled up an incline and allowed to roll back. Make sure the can is always at least 0.4 meters away from the sonic ranger. Predict what the shape of distance vs. time and velocity vs. time graphs will look like for the can or ball rolled up and down the incline.

   7. Make a sketch of your predicted graphs.

Fig. A

Select the option so that both distance vs. time and velocity vs. time are displayed simultaneously.

8. Sketch the shape of the distance vs. time graph and velocity vs. time graphs.

9. Is the velocity vs. time graph a straight line?

Part C: Analyze Pendulum Motion

Fig. B

Step 9: Use the sonic ranger to analyze the motion of a pendulum as in Figure B. The sonic ranger needs to always be a minimum of 0.4 meters from the swinging bob. Use a pendulum with a length great enough (such as a meter or longer) so that the path of the bob is nearly a straight line in the beam of the sonic ranger. Collect data of the swinging pendulum, then choose the option so that both distance vs. time and velocity vs. time graphs are displayed simultaneously.

10. Sketch the shape of the two graphs as they are displayed on the monitor.

11. Where is the speed greatest for the swinging pendulum bob? Least?

Blind as a Bat

Part D: Analyzing the Acceleration of Gravity

Step 10: Place the sonic ranger on the floor and stack a few books around to protect it. Then drop large heavy objects (so that the effect of air friction is negligible) like heavy beach balls directly on it.

12. Describe the distance vs. time and velocity vs. time graphs. How do they compare?

_____

_____

_____

Summing Up

13. Is the acceleration constant (or nearly so)?

_____

_____

14. How does this compare to the can rolling down the incline in Part B?

_____

_____

_____

_____

CONCEPTUAL **Physics** **Experiment**

Projectile Motion

On Target

Purpose
To predict the landing point of a projectile.

Required Equipment and Supplies
ramp or Hot Wheels track
1/2" (or larger) steel ball
empty soup can
meterstick
plumb line
stopwatch or
LabTools and *Bull's Eye* software or equivalent
computer
interface box
light probes

Discussion
If you were to toss a rock in some region of gravity free outer space, it would just keep going — indefinitely. The rock would continue its motion at constant speed and cover a constant distance each second (Figure A). When motion is uniform, the equation for distance traveled is

Fig. A

$$d = \bar{v}t$$

and the speed is

$$\bar{v} = \frac{x}{t}$$

Back on earth, what happens when you drop a rock? It falls to the ground and the distance it covers in each second increases (Figure B). Gravitation is constantly increasing the speed of the rock. If we let y represent vertical distance (and x horizontal distance) then the equation of the vertical distance fallen in t seconds is:

Fig. B

$$y = \tfrac{1}{2}gt^2$$

where g is the acceleration due to gravity. Starting from rest, the instantaneous falling speed v after time t is

$$v = gt$$

What happens when you toss the rock horizontally (Figure C)? The curved motion that results can be described as the combination of two straight-line components of motion: one vertical and the other horizontal. The vertical component undergoes acceleration, while the horizontal component does not. The secret to analyzing projectile motion is to keep two separate sets of "books": one that treats the horizontal motion according to

$$x = \bar{v}t$$

and the other that treats the vertical motion according to

$$y = \tfrac{1}{2}gt^2$$

On Target

Fig. C

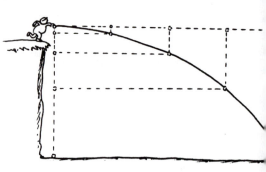

Horizontal Motion
- When thinking about how *far*, think $y = \frac{1}{2}gt^2$
- When thinking about how *fast*, think $\bar{v} = \dfrac{x}{t}$

Vertical Motion
- When thinking about how *far*, think $y = \frac{1}{2}gt^2$
- When thinking about how *fast*, think $v = gt$

When engineers build bridges or skyscrapers they do *not* do so by trial and error. For the sake of safety and economy, it must be right the *first* time. Computer simulations enable engineers to check and double check their calculations. Your goal in this experiment is to predict where a steel ball will land when released from a certain height on an incline. The final test of your measurements and computations will be to position an empty soup can so that the ball lands in the can on the *first* attempt.

If available, the *Bull's Eye* computer simulation will allow you to enter your measurements and test your prediction on the computer before you actually run the experiment — thereby eliminating costly mistakes. You should do the same!   **Fig. D**

Procedure
Step 1: Assemble your ramp. Make it as sturdy as possible so the steel balls roll smoothly and reproducibly, as shown in Figure D. The ramp should not sway or bend. The ball must leave the table *horizontally*. Make the horizontal part of the ramp at the bottom at least 10 cm long. The vertical height of the ramp should be at least 40 cm.

Step 2: Measure the time it takes the ball to travel from the first moment it reaches the level of the table top (point A in Figure D) to the time it leaves the table top (point B in Figure D). Divide the horizontal distance on the ramp (from point A to point B) by this time interval into to find the horizontal speed. Release the ball from the same height (marked with tape) on the ramp several times so that your timings are consistent. Do *not* cheat and allow the ball to strike the floor! Record the average horizontal speed.

    horizontal speed = _____ cm/s

Step 3: Using a plumb line and a string, measure the vertical distance *y* the ball must drop from the bottom end of the ramp in order to land in an empty soup can on the floor.

    1. Should the height of the can be taken into account when measuring the vertical distance *y*? If so, make your measurements accordingly.

_____

Step 4: Using the appropriate equation from the discussion, find the time *t* it takes the ball to fall from the bottom end of the ramp to the can. Write the equation that relates *y* and *t*.

    equation for vertical distance: _____

Step 5: The range is the horizontal distance a projectile travels, *x*. Predict the range of the ball. Write the equation you used to predict the range. Write down your predicted range.

On Target

equation for the range: _____

predicted range $x =$ _____

$$y = \frac{1}{2}gt^2$$

$$v = gt$$

Now place the can on the floor where you predict it will catch the ball.

Step 6: After your instructor has checked your predicted range and your can placement, release the ball from the marked point on the ramp.

Summing Up

2. Did the ball land in the can?

_____

3. What may cause the ball to miss the target?

_____

4. You probably noticed that the range of the ball increased in direct proportion to the speed at which it left the ramp. The speed depends on the release point of the ball on the ramp. What role do you think air resistance had in this experiment?

_____

5. Briefly discuss your percentage error and in what direction is the error most likely to be and why.

_____

_____

_____

CONCEPTUAL **Physics**

Discovering Relationships

Trial and Error

Purpose
To discover Kepler's third law of planetary motion through a procedure of trial and error using the computer.

Required Equipment and Supplies
Data Plotter software or equivalent
computer

Discussion
Pretend you are a budding astronomer. In order to earn your Ph.D. degree, you are doing research on planetary motion. You are looking for a relationship between the time it takes a planet to orbit the sun (its *period*) and the average radial distance of the planet's orbit around the sun. It is customary to express radial distances in *AU*s, Astronomical *Units*, where 1 AU is the average radius of the earth's orbit. Using a telescope, you have accumulated the planetary data shown in Table A.

You have access to a computer program that allows you to not only plot data easily, but also to plot many different relationships between the variables that make up your x and y coordinates. For example, in addition to being able to plot period T vs. radius R, you can also plot T^2 vs. R, T vs. R^2, T vs. R^3, and so on. To discover how T and R are related, you must find the combination of powers which results in a graph which is a straight line. A *linear* graph means the quantity you are plotting on the vertical axis is *directly proportinal* to the quantity you are plotting on the horizontal axis. That is, doubling x doubles y, tripling x triples y, and so on. Hence, the relationship between the variables is simply the *ratio* of the variables raised to the powers whose graph is a straight line. For example, if T vs. R^2 is a straight line, then $T \sim R^2$, or $T/R^2 \sim$ constant.

Suppose you are plotting T vs. R. The name of the game is to get a straight line. That's because a straight line tells you that whatever you are plotting on the y-axis is *proportional* to whatever you are plotting on the x-axis. If, however, your graph curves upward as in Figure A, it means T is increasing *faster* than R. That suggests you should try *increasing* the power of the x-values (R).

On the other hand suppose your graph of T vs. R looks like Figure B. That means R is increasing *faster* than T. Straighten-out the graph by re-plotting with a higher power of the y-values (T).

Procedure
Step 1: Follow the instructions on the *Data Plotter* program. Input the data from Table A. Select the "Graph Set-Up" option to vary the powers of the x and y values. Sometimes the up or downward curve of a graph is *very* subtle to observe on a computer monitor, and a curved graph can easily be mistaken for a straight line. The "Least Squares Fit" option is helpful because it plots the best average straight line of your data. This enables you to see much more clearly if your graph is curving up or down.

Table A

PLANET	PERIOD (YEARS)	AVERAGE RADIUS (AU)
MERCURY	0.241	0.39
VENUS	0.615	0.72
EARTH	1.00	1.00
MARS	1.88	1.52
JUPITER	11.8	5.20
SATURN	29.5	9.54
URANUS	84.0	19.18
NEPTUNE	165.	30.06
PLUTO	248.	39.44

Fig. A

Fig. B

Step 2: If the relationship between T and R cannot easily be discovered by modifying the power of only one variable at a time, try modifying the power of both variables at the same time.

1. How are T and R related?

_____

_____

_____

_____

If you find the relationship between T and R during this lab period, feel *good*. It took Johannes Kepler (1561-1630) ten *years* of painstaking effort to discover the relationship. Computers were not around in the 16th century!

Going Further

After modifying the powers of both variables, try re-plotting your data (restore your data to the first power) by graphing the logarithm (either base *e* or base 10 is OK) of the period vs. the logarithm of the average radius. Describe the resulting graph. What is the slope of the graph? How is this related to the functional relationship between the period and the average radius?

Analyzing your data in this manner is a powerful skill. Try performing the same procedure on other sets of data such as those in Table B or others supplied by your instructor. What relationships, if any, do you discover?

Table B

DISTANCE (cm)	INTENSITY (%)
30	100
35	73
40	56
42.4	50
45	44.4
50	36
55	29
60	25
65	21
70	18
75	16
90	11.1
105	8.2
120	6.25

Trial and Error

CONCEPTUAL Physics ━━━━━━━━━━━━━━━━━━━━━━━━━━━━━
Balanced Torques

Weight a Moment

Purpose
To apply the principles of torques to a simple lever.

Required Equipment and Supplies
meterstick
wedge or knife-edge
two 50-g mass-hangers
slotted masses or set of hook masses
knife-edge lever clamps
masking tape
rocks with masses of approximately 100 g
triple beam balance
bathroom scale

Discussion
The see-saw is a simple mechanical device that rotates about a pivot (a *fulcrum*). It is a type of lever. Although the work output by a device can never be more than the work or energy input, levers make it *easier* to accomplish a variety of tasks.

Suppose you are an animal trainer at the circus. You have a very strong, very light, wooden plank. You want to balance a 6300 kg. elephant on a see-saw using only your own body weight. Suppose your body has a mass of 50 kg. The elephant is to stand 2 meters from the fulcrum. How long must your plank be in order for you to balance the elephant by standing at its far end?

Laboratories do not normally have elephants or masses that size. They do have a variety of smaller masses, metersticks, and fulcrums that enable you to discover how levers work, describe their forces and torques mathematically, and finally solve the elephant problem.

Procedure
Step 1: Carefully balance a meterstick horizontally on a wedge or knife edge. Suspend a 200-g mass 10 cm from the fulcrum. Suspend a 100-g mass on the opposite side of the fulcrum at the point that balances the meterstick. Record the masses and distances from the fulcrum in Table A.

1. Can a heavier mass be balanced by a lighter one? Explain how.

_____

_____

_____

_____

Step 2: Make more trials to fill in Table A. You can do this by using the masses of Step 1 and changing their positions. For instance, you can move the heavier mass to a new location 5 cm farther away, and then rebalance the meterstick with the lighter mass.

Table A

TRIAL	SMALL MASS (g)	DISTANCE FROM FULCRUM (cm)	LARGE MASS (g)	DISTANCE FROM FULCRUM (cm)

You can also change the masses. Replace the heavier mass with another mass and rebalance the lever by moving the lighter mass. Record the masses and distances from the fulcrum in Table A. Be sure to include the mass of any hanger or clamp.

Step 3: Use any method you can devise to discover a pattern in the data of Table A. You can try graphing the large mass vs. its distance from the fulcrum, the small mass vs. its distance from the fulcrum, or another pair of variables. You can also try forming ratios or products to discover the pattern.

Step 4: After you have convinced yourself and your partner that you have discovered a pattern, convert this pattern into a word statement.

_____

_____

_____

_____

HMM... I WONDER IF...

Weight a Moment

Step 5: Now convert this word statement into a mathematical equation. Explain what each symbol represents.

_____

_____

Step 6: With the help of your partners or your instructor, use your equation to find the distance a 50-kg person should stand from the fulcrum in order to balance the elephant. Show your calculations.

Answer: $d =$ _____ m

Summing Up

2. Why must the mass of the hangers and clamps be taken into account in this experiment?

_____

_____

3. Suppose you are playing see-saw with your younger sister who weighs much _less_ than you. What can you do to balance the see-saw? Mention two things.

_____

_____

Weight a Moment

CONCEPTUAL Physics ──────────────────────────────────── | Experiment |

CG and Balanced Torques

Solitary See-Saw

Purpose

To use the principle of balanced torques to find the value of an unknown mass.

Required Equipment and Supplies

meterstick
2 mass clamps for meterstick
set of slotted masses
2 mass hangers
fulcrum
fulcrum holder
string

Discussion

Gravity pulls on every part of an object. It pulls more strongly on the more massive part of objects and more weakly on the less massive parts. The sum of all these pulls is the weight of the object. The average position of the weight of an object is its center of gravity, CG (Chapter 7 in *Conceptual Physics*).

The whole weight of the object is effectively concentrated at its CG. The CG of a uniform meterstick is at the 50-cm mark. In this experiment you will balance a meterstick with a known mass, and compute the mass of the unknown using the balanced torque equation, $F_1 d_1 = F_2 d_2$. Then you will simulate a "solitary see-saw" by balancing the weight of a meterstick with a known weight .

Procedure

Step 1: Balance the meterstick horizontally with nothing hanging from it. Record the position of the CG of the meterstick.

position of the meterstick CG = _____

Using a string, attach an object of unknown mass, such as a rock, at the 90 cm mark of the meterstick, as shown in Figure A. Place a known mass on the other side to balance the meterstick. Record the known mass and its position.

Fig. A

mass = _____   position = _____

Step 2: Measure the distances from the fulcrum to each mass.

d_1 = distance from fulcrum to known mass = _____

d_2 = distance from fulcrum to unknown mass = _____

These two distances are known as *lever arms*. The lever arm is the (perpendicular) distance from the fulcrum to the place where the force acts. Write down the equation for the balanced torques with the known values. Calculate the unknown mass.

equation: _____

mass$_{calculated}$, m = _____

Step 3: Measure the unknown mass using a balance or spring scale.

1. How does your calculated value compare with your measured value for the unknown mass?

_____

_____

Step 4: Place the fulcrum exactly on the 85-cm mark. Balance the meterstick using a single known mass that you hang between the 90-cm and 95-cm mark as in Figure B. Record the mass used and its position. **Fig. B**

known mass = _____ g   position = _____ cm

Step 5: Draw a lever diagram of your meterstick system. Be sure to label the fulcrum, the masses giving rise to torques on each side of the fulcrum, and the lever arm distance for each mass.

2. Where is the entire mass of the meterstick *effectively* located?

_____

Step 6: Use the balanced torque equation to calculate the mass of the meterstick.

mass$_{calculated}$, m = _____ g

Step 7: Remove the meterstick and its knife-edge. Find its mass using a balance.

mass $_{measured}$, m = _____ g

Summing Up
3. How do the values for the mass of the meterstick compare? Calculate the percentage difference (see the Appendix on how to do this if necessary).

_____

_____

CONCEPTUAL **Physics**   ■ **Experiment**

Leverage and MA

Gearing Up

Purpose
To calculate the mechanical advantages of the gears on a ten-speed bicycle.

Required Equipment and Supplies
ten-speed bicycle
meterstick

Discussion
When a simple lever is balanced, the sum of the torques on one side of the fulcrum are equal and opposite to the sum of the torques on the other side. This lever principle or torque equation (Chapter 7 in *Conceptual Physics*) can be summarized as

$$F_1 d_1 = F_2 d_2$$

In the "Weight a Moment" lab, you learned that you can solve for any one variable if the other three are known. Levers are usually designed with a specific purpose in mind. Although levers can't reduce the amount of work to be done, they can make it easier. *Mechanical advantage* (MA) is a measure of this reduction in effort.

Fig. A Force
More torque

Lever arm

Mechanical advantage can be advantageous in two different ways. A smaller effort force positioned farther from the fulcrum of a lever can overcome a larger resistance force. The price paid is moving the smaller force through a larger distance. Figure A shows a lever which results in such a gain in *force*. The lever in Figure B results in a gain in *distance* or *speed*. A larger effort force balances a smaller resistance force a greater distance from the fulcrum. In either case, the *energy output* can never be greater than the *energy input*.

Fig. B Force
Torque

Lever arm

Procedure
Step 1: When pedaling the bike, the tire pushes against the road and the road pushes against the tire. Which of these two forces acts on the tire (and therefore the bike) and moves it down the road?

_____

Analyze in a qualitative sense (without using any specific measurements) whether the force exerted by the tire on the road is greater or less than the weight of the rider. Do this by examining the lever system consisting of the foot pedal, the front sprocket, the rear sprocket, and the rear wheel. (The chain only transmits the force from the front gear to the rear gear; compare this to the activity "Tug of War".) Investigate a variety of gear combinations. Which gear combination results in the *greatest* force?

_____

Which results in the *smallest* force?

_____

Step 2: Now let's quantitatively (by making specific measurements) compute the force the tire exerts against the *road* (and hence the road exerts against the *tire*) when you push against the *pedal*. Set the pedal in the horizontal position, as shown in Figure C. Assume the force you exert on the pedal is your own weight (as it would be if you stood straight up on the pedal). Measure the four lever arms involved.

Fig. C

If you have trouble seeing how to do this, study Figure D. To simplify, use the following variables for the lever arms.

d_1 = lever arm for front pedal
d_2 = lever arm for front sprocket
d_3 = lever arm for rear sprocket
d_4 = lever arm for rear wheel

Fig. D

and the following variables for the forces

F_1 = your weight
F_2 = force exerted on the chain
F_3 = force exerted by the chain
F_4 = force of the tire against the road

At the instant you step on the pedal, the torques on the pedal and the front sprocket system are equal to one another.

$$F_1 d_1 = F_2 d_2$$

The force exerted on the chain by the front sprocket is transmitted directly to the rear sprocket.

$$F_2 = F_3$$

The torque from the rear sprocket on the axle is equal to the torque on the wheel from the road.

$$F_3 d_3 = F_4 d_4$$

Step 3: Measure the lever arms involved when the bike is in low (1st) gear.

d_1 = lever arm for front pedal = _____
d_2 = lever arm for front sprocket = _____
d_3 = lever arm for rear sprocket = _____
d_4 = lever arm for rear wheel = _____

Compute the force the tire exerts against the road.

F_{road} = _____

Step 4: Repeat for the bike when it is in high (10th) gear.

d_1 = lever arm for front pedal = _____
d_2 = lever arm for front sprocket = _____
d_3 = lever arm for rear sprocket = _____
d_4 = lever arm for rear wheel = _____

F_{road} = _____

Gearing Up

Summing Up

1. What is the best combination of gears that results in the maximum *force* on the rear wheel?

_____

_____

2. When pedaling a bike in low gear, what price must be paid in terms of the effort force or distance, and what benefit is received?

_____

_____

3. What is the best combination of gears for the greatest *distance* or *speed*?

_____

_____

4. When pedaling a bike in high gear, what price must be paid in terms of the effort force or distance, and what benefit is received?

_____

_____

CONCEPTUAL **Physics** | **Activity**

Density

Get the Lead Out

Purpose
To observe the result when objects of different densities are placed in water.

Required Equipment and Supplies
blocks of lead, wood, and Styrofoam of equal mass
variety of soda pop in aluminum cans; both diet and regular varieties
aquarium tank or sink
egg
wide-mouth graduated cylinder
bowl
salt
spoon
balance

Discussion
Have you ever noticed how some people have trouble floating when
swimming while others don't? This activity will help you understand
why.

Procedure
Step 1: Using a double pan balance, balance equal masses of lead and
wood. Repeat, using an equal mass of Styrofoam.

1. How do the *volumes* compare? How do the *densities* compare?

_____

_____

Step 2: Try floating a variety of cans of soda pop.

2. Which ones float? Sink?

_____

_____

3. How does the density of the different kinds of soda pop compare to the density of
tap water?

_____

_____

4. How do the relative densities relate to sugar content?

_____

_____

Step 3: Use a balance to measure the mass of an egg. Using a wide mouth graduated cylinder, carefully determine the volume of the egg by measuring the volume of water it displaces when it is slowly (gently!) lowered into the graduated cylinder. Calculate its density: $d = m/V$.

density = _____

Step 4: Now try to float the egg in a bowl of water. Does it float? If not, dissolve enough salt in the water until the egg floats.

5. How does the density of an egg compare to that of tap water?

_____

_____

6. To salt water?

_____

_____

Summing Up

7. Does adding salt to the water make the water less dense or more dense? How?

_____

_____

_____

_____

_____

8. Why do some people have trouble floating when swimming while other don't?

_____

_____

Get the Lead Out

CONCEPTUAL Physics

Scaling

Activity

Elephant Ears

Purpose
To investigate surface area to volume ratios.

Required Equipment and Supplies
Styrofoam balls of diameter 0.5, 1, 2, and 4 inches

Discussion
Why do elephants have big ears? Why do large creatures like elephants eat less food per body weight than small creatures like mice? Why is falling dangerous for large creatures and relatively harmless for small ones? These questions have to do with scaling (Chapter 11 of your text) and the geometrical relationships between the surface areas and volumes of things, whether they are living creatures or inanimate objects. The purpose of this activity is to investigate ratios of surface to volume.

Procedure
Drop pairs of different size Styrofoam balls from a height of about 3 m. Hold the balls so that the bottom of the smaller ball is even with the bottom of the larger ball, and be sure to release the balls at the same instant. Compare their falling times by simply observing which ball strikes the ground first, without going to the bother of measuring their actual falling times. Summarize your observations in Table A.

1. Describe any pattern you observe.

_____

2. Which size Styrofoam ball had the greatest average speed when falling?

_____

3. Which size Styrofoam ball has the greatest surface area?

_____

4. Which size Styrofoam ball has the greatest volume?

_____

5. Which size Styrofoam ball has the greatest ratio of surface area to volume?

_____

_____

Table A

FALLING TIME	BALL DIAMETER
HITS THE GROUND 1ST	
HITS THE GROUND 2ND	
HITS THE GROUND 3RD	
HITS THE GROUND LAST	

Summing Up

A Styrofoam ball falling through the air encounters two principal forces. One is the downward force due to gravity — its weight, which for uniform objects such as a Styrofoam ball, is proportional to its volume. The other force is the upward force of air resistance — *drag* — which opposes its fall. Drag depends on the surface area of the ball.

6. Which size Styrofoam ball encounters the greatest drag?

_____

7. Which size Styrofoam ball encounters the greatest *net* force (due to gravity *and* drag)?

_____

8. Are your answers to these questions consistent with your observations?

_____

9. Based on this experience, predict which will fall to the ground faster — a large raindrop or small raindrop. Why?

_____

_____

_____

Elephant Ears

CONCEPTUAL Physics ─────────────────────────────────┐ **Activity**

Pressure and Force

Strong as an Ox

Purpose
To compute the force holding two Magdeburg disks together.

Required Equipment and Supplies
Magdeburg disks
vacuum pump

Discussion
The famous "Magdeburg hemispheres" experiment of 1654 demonstrated the enormous strength of atmospheric pressure. Two teams of horses couldn't pull them apart. Why were the hemispheres held together? By what? In this activity, you'll experience it yourself!

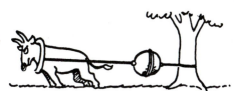

Procedure
Your instructor will evacuate a pair of Magdeburg hemispheres for you (see Figure 13.2 in your text). Team up with another student and see if you can pull the hemispheres apart. *Caution:* Have other students brace to catch you in the event the hemispheres suddenly break apart due to a poor seal — so that you don't lose your balance and get hurt! Are you and your fellow student able to pull them apart?

Now, compute the force necessary to separate the hemispheres.

$$\text{pressure} = \frac{\text{force}}{\text{area}}$$

$$\text{force} = \text{pressure} \cdot \text{area}$$

Measure the diameter of a hemisphere and compute the cross-sectional area in square inches:

$$\text{area} = \pi r^2$$
$$= \underline{\hspace{2cm}} \ \text{in}^2$$

Assume air pressure is 14.7 lb/in^2 (equivalently 10^5 N/m^2) and for simplicity that the disks are completely evacuated to a perfect vacuum.

$$\text{force} = 14.7 \ \text{lb/in}^2 \cdot (\text{area}) \ \text{in}^2$$

$$= \underline{\hspace{2cm}} \ \text{lb}$$

Now do you understand why pulling the disks apart is so difficult?

CONCEPTUAL Physics   **Activity**

Elasticity

Visco-Elasticity of Corn Starch

Purpose
To explore an unusual property of a common substance.

Required Equipment and Supplies
corn starch
2" rubber mallet
pie pan
water

Discussion
Toys known as Silly Putty and Slime are *visco elastic*. When pulled
slowly they stretch; when jerked, they break in two; when rolled into a
ball, they bounce on a hard floor. A common example of a visco-elastic
material is a mixture of cornstarch and water.

Procedure
Pour about half of a box of corn starch into a sturdy pie tin. Thoroughly
mix in enough water so that it has the consistency of a thick milkshake.
Put your bare hand in it and play around with it. Lift some and allow it
to slowly drip into the pie tin to ascertain its liquid properties. Then
strike the mixture with a rubber mallet.

How does the corn starch behave — like a liquid or like a solid? Does it splatter? How
would you describe the state of matter of the corn starch?

_____

_____

_____

_____

For interesting reading, check out the *The Amateur Scientist* article entitled "Why do
honey and syrup form a coil when they are poured?" by Jearl Walker, in the September
issue of *Scientific American*, (Volume 245, #3, p. 216-226), 1981.

CONCEPTUAL **Physics** ———————————— | **Activity** |

Polarity of Molecules

Polarity of Molecules

Purpose
To investigate the effect of a charged rod on tiny streams of liquids.

Required Equipment and Supplies
sewing needle
buret and support
water
triclorotriflouroethane (TTE)

Discussion
A water molecule consists of one atom of oxygen and two atoms of hydrogen (Figure 21.14 in *Conceptual Physics*). Although the water molecules themselves have no net electrical charge, there is a slight displacement of positive and negative charges in the water molecule. This is because the oxygen nucleus has a greater electric charge than either hydrogen nucleus, and the electrons of the two hydrogen atoms are more attracted to the oxygen nucleus and spend more time near it than either of the two hydrogen atoms. We say the water molecule is electrically *polarized*. More specifically, the opposite charges on opposite ends create what is called an electric *dipole*. The oxygen end of the molecule behaves as if it were slightly negatively charged and the hydrogen end behaves as if it were slightly positively charged.

The polarity of water molecules accounts for some of water's unique properties. The positive end of one water molecule is attracted to the negative end of another to form weak bonds. In the solid state, the atoms group in such a way as to create a hexagonal crystal lattice that occupies more space than the individual unbonded atoms. So ice floats. When the temperature increases and the ice melts to form water, the weak attraction between water molecules causes *surface tension*, making it possible to float needles and razor blades. As the temperature increases, the molecules are not so able to form bonds and surface tension decreases.

Procedure
Step 1: Carefully float a needle on the surface of water at room temperature. Heat the water and see what happens. Explain.

_____

_____

Step 2: Briskly rub a piece of cat's fur on a hard rubber rod to charge the rod negatively. Bring the charged rod near a stream of water from a buret prepared by your instructor. What happens to the stream of water? Why?

_____

_____

Step 3: What do you think would happen if you used a positively charged rod? Does the stream behave differently than before? Try it and see.

_____

_____

Step 4: Repeat using a non-polar substance such as triclorotrifluoroethane (TTE). How do your results compare?

_____

_____

Summing Up
Make a sketch of how the charges in the falling water arrange themselves in the presence of a negatively charged rod. Make another sketch for a positively charged rod.

CONCEPTUAL Physics ━━━━━━━━━━━━━━━━━━━━ | **Activity**

Fluid Pressure

Screwball Bernoulli

Purpose
To observe and analyze some everyday circumstances involving Bernoulli's principle.

Required Equipment and Supplies
hair dryer
Ping Pong ball
meter-long tube

Discussion
Did you know that windows of a house in a tornado are actually blown out by the air inside the house? How can that be? Rats to you, Daniel Bernoulli!

Procedure
Use a hairdryer on its high speed setting and blow upward on a Ping Pong ball. Keep the ball as steadily as possible in the vertical air stream. Slowly tilt the hairdryer sideways and see how far you can go before the ball falls. Make a sketch of the streamlines of the air around the ball in a vertical orientation and in a non vertical orientation.

Place a Ping Pong ball in one end of a plastic tube that has a diameter slightly larger than that of the ball. Whip the tube quickly in a downwards motion so that the ball rolls out of the tube in the forward direction. It may take a few times to perfect your technique. Observe the motion of the ball. How does it curve after it leaves the tube?

_____

Repeat whipping the tube, this time sideways. Have your partner pretend to be a batter. Which way does the ball curve?

_____

Which way does the ball have to spin to curve away from the batter (a slider)?

_____

Towards the batter (a screwball)?

_____

Summing Up
Make a sketch of the streamlines around the ball for these two cases. Using Bernoulli's principle, explain why the ball curves in the direction it does.

_____

_____

_____

Motion of air relative to ball

CONCEPTUAL **Physics** | **Activity**

Displacement by Volume

Get Out of the Way!

Purpose
To show that the mass of water displaced for a submerged object depends on the volume of the submerged object and not the mass or weight.

Required Equipment and Supplies
35 mm film canisters
string
ballast material (nuts, BB's, sand, etc.)
beam balance
500 mL graduated cylinders

Discussion
Archimedes' is very famous because he discovered a clever way to measure the volume of an irregularly-shaped object — the king's crown. His discovery is based on a very simple idea that many people don't fully understand. After doing this activity, you won't be one of them!

Procedure
Your instructor will provide you with two film canisters with a piece of string attached. Find the mass of each canister.

mass of lighter canister, m = _____ g

mass of heavier canister, m = _____ g

Procedure
Attach a strip of masking tape vertically at the water level of a graduated cylinder about $\frac{3}{4}$ full of water. Mark the water level on the tape. Submerge the lighter canister and mark the new water level on the tape. Remove the canister.

Predict what the water level for the heavier canister will be when it is submerged?

_____

Try it and see! Was your prediction correct?

_____

Summing Up
Explain your observations.

_____

_____

_____

CONCEPTUAL **Physics** | **Activity** |

Buoyant Force

Cartesian Diver

Purpose
To show that the buoyant force on an object depends on the amount of fluid displaced.

Equipment and Supplies:
empty 2 liter plastic soda pop container
2 mL plain dropping pipet (eye-dropper)
food coloring (any bright but clear color such as orange)

Discussion
How does a fish swim to the bottom of a pond? How does a submarine dive to the depths of the ocean? In this activity you will build a simple device that will not only help you understand Archimedes' principle, but to *see* just how it actually works.

Procedure
Make a Cartesian diver by filling a 2 liter plastic soda pop container with colored water. Then place an eye dropper (2 mL dropping pipette works fine) which is about one third full of water in the container. Adjust the amount of water in the eye dropper so that it just floats. Top off the container with water and screw the cap on tightly so that it is airtight.

Now squeeze the container and watch the eye dropper dive! Observe what happens to the water level (bubble) inside the eye dropper as it dives. Describe your observations.

_____

_____

_____

Summing Up
1. Does the volume of water displaced by the air bubble increase or decrease when the container is squeezed?

_____

2. How does this affect the buoyant force?

_____

_____

3. Can you offer an explanation for the behavior of the diver?

_____

_____

Cartesian Diver

CONCEPTUAL Physics ████████████████████████████ **Experiment**

Force and Pressure

Treading Air

Purpose
To distinguish between pressure and force.

Required Equipment and Supplies
automobile
owner's manual for vehicle (optional)
graph paper
tire pressure gauge

Discussion
People commonly confuse *force* and *pressure*. Tire manufacturers add
to this confusion by saying "Inflate to 45 pounds" when they really
mean "45 pounds *per* square inch." The pressure on your feet is
painfully more when you stand on your toes than when you stand on
your whole feet, even though the force of gravity (your weight) stays
constant. The pressure depends on how a force is distributed over a
certain area. Pressure decreases as the area increases.

$$\text{pressure} = \frac{\text{force}}{\text{area}}$$

$$\text{force} = \text{pressure} \times \text{area}$$

Procedure
Step 1: Position a piece of graph paper directly in front of each front tire of an
automobile. Roll the automobile onto the paper.

Step 2: Use a tire gauge to measure the pressure in each tire in pounds per square
inch. Write down the pressures of each tire on the graph paper. Trace the outline of
the tire where it makes contact with the graph paper. Roll the vehicle off the paper.
Calculate the area inside the trace in square inches or square cm. Record your data in
Table A.

Step 3: Study the tread pattern of the tire. Compare this to the tire mark on the
paper. Note that only the tread actually presses against the road. The gaps between
the tread do not support the vehicle. Estimate what percentage of the tire's area of
contact is tread and record it in Table A. The area actually in contact with the road is
the area inside your trace multiplied by the percentage of tread. Repeat for all four
tires.

Step 4: Compute the force each tire exerts against the road by
multiplying the pressure of the tire times its area of contact with the
road. Record your computations in Table A.

Step 5: Compute the weight of the vehicle by adding the forces from
each tire.

computed weight = _____ lb

Table A

TIRE	OUTLINE TRACED ON PAPER (in²)	ESTIMATE OF % TREAD	AREA OF CONTACT (in²)	PRESSURE (lb/in²)	FORCE (lb)
RIGHT FRONT					
LEFT FRONT					
RIGHT REAR					
LEFT REAR					

Step 6: Ascertain the weight of the vehicle from another source, such as the owner's manual or the local dealer. Sometimes this information is stamped on a plate on the inside of a door jamb.

known weight = _____ lb

Summing Up

1. How does your computed value of vehicle weight compare to the stated value?

_____

2. How could you make your estimate of the percentage of tread more accurate?

_____

3. What could account for a difference in the actual weight of a car and the weight stated in the owner's manual?

_____

4. In terms of pressure, why is it important to have good tread on your tires, particularly in rainy weather?

_____

5. A tire gauge measures the difference between the pressure in the tire and atmospheric pressure (14.7 lb/in² or 10^5 N/m²) outside the tire. What is the *total* pressure inside the front tire?

_____

6. Why was the atmospheric pressure (14.7 lb/in²) *not* added to the pressure of the tire gauge when computing the weight of the vehicle?

_____

_____

7. Why do trucks that carry heavy loads have so many wheels — eighteen in some cases?

_____

_____

CONCEPTUAL **Physics**   ███ **Experiment** ███

Scaling

Hot Pots

Purpose
To investigate how (surface area/volume) ratio affects the rate of cooling.

Required Equipment and Supplies
50-mL beaker
400-mL beaker
1000-mL beaker
stopwatch or clock with second hand
hot plate
3 thermometers,
   or
LabTools software or equivalent
computer
interface box
3 temperature probes
printer

Discussion
Which cools a soda faster — ice cubes or crushed ice? Which heats up
faster in the hot sun — a swimming pool or a wading pool? Which cools
quicker when the sun goes down? Suppose you had two similar pots,
one with several times the volume of the other, filled with hot water.
Which pot of water would cool down faster?

CAUTION
BE EXTRA CAUTIOUS
WITH BOILING
WATER AND STEAM!

The surface of an object is altogether different than the volume of an
object (Chapter 11 of your text). Cooling takes place at the surface of an
object. The molecules of air near the surface are warmed by the liquid,
rise, and are replaced by cooler air. Thus, the rate of cooling depends on
the surface area. As you will see, the cooling rate also depends on the
volume.

Caution: Handling boiling hot water can be dangerous! Always use extreme care.

Procedure
Heat 1500 mL of water to about 80°C. *Carefully* pour 1000 mL, 400 mL, and 50 mL
into their respective beakers in that order. Allow the beakers to cool down by
themselves on the lab table. Measure the temperature every 30 seconds if you are
using a temperature probes connected to a computer and every minute if you are using
thermometers. If you are using a computer, make sure the probes are properly
calibrated before you begin. Either record your data in Table A or printout a copy of
your graph from the computer.

Hot Pots

Summing Up

1. Which beaker of warm water cooled fastest?

_____

2. Which beaker has the most surface area?

_____

3. Which beaker has the greatest volume?

_____

4. Which beaker has the largest ratio of surface area to volume?

_____

5. The cooling of a beaker of warm water takes place mainly at the surface. The total amount of heat that dissipates from the water for it to cool to room temperature depends on the volume. Therefore, on what does the *rate* of cooling (the temperature drop with time) depend?

_____

_____

_____

Table A

TIME (s)	TEMPERATURE	
	50 mL	400 mL
0		
30		
60		
90		
120		
150		
180		
210		
240		
270		
300		

6. Suppose that a small wading pool is next to a swimming pool. Predict which pool will heat up faster during the day and cool down slower during the night? Why?

_____

_____

CONCEPTUAL Physics

Hooke's Law

Experiment

By Hooke or By Crook

Purpose
To verify Hooke's law and determine the spring constants for a spring and a rubber band.

Required Equipment and Supplies
ring stand or other support with rod and clamp
3 springs
paper clip
masking tape
meterstick
set of slotted masses
large rubber band
graph paper or graphing program and computer

Discussion

Fig. A

When a force is applied to an object, the object may be stretched, compressed, bent, or twisted. The electrical forces between atoms in the object resist these changes. When the applied force is removed, the electrical forces return the object to its original shape. Too large an applied force may overcome these resisting forces and cause the object to deform permanently. The minimum amount of force that permanently deforms the object is called the *elastic limit*.

Hooke's law (page 198 in your text) applies to changes below the elastic limit. It states that the amount of stretch or compression is directly proportional to the applied force. The proportionality constant is the *spring constant, k*. Hooke's law is written $F = kx$, where F is the applied force and x is the distance moved. A stiff spring has a high spring constant and a weak spring has a relatively smaller spring constant.

Procedure

Step 1: Hang a spring from a support. Attach a paper clip to the free end of the spring with masking tape. Clamp a meterstick in a vertical position next to the spring (Figure A). Note the position of the bottom of the paper clip relative to the meterstick. Place a piece of masking tape on the meterstick with its lower edge at this position.

Step 2: Attach different masses to the paper clip at the end of the spring. With your eye at the same level with the bottom of the paper clip, note its position each time. The stretch in each case is the difference between the positions of the paper clip when a load is on the spring and when no load is on the spring. Be careful not to exceed the elastic limit of the spring. Record the mass and corresponding stretch of each trial in the first row of Table A.

Step 3: Repeat Steps 1 and 2 for two different springs. Record the masses and stretches in the 2nd and 3rd rows of Table A.

Step 4: Repeat Steps 1 and 2 using a large rubber band. Record the masses and stretches in Table A.

Step 5: Calculate the weight of each mass, and record it in Table A. On graph paper, make a graph of force (vertical axis) vs. stretch (horizontal axis) for each spring and the rubber band. If a graphing program such as *Data Plotter* is available, enter and plot your data using a computer.

Step 6: For each graph that is an upward sloping straight line, draw a horizontal line through one of the lowest points on the graph. Then draw a vertical line through one of the highest points on the graph. Now you have a triangle. The slope of the graph equals the vertical side of the triangle divided by the horizontal side. The slope of a Force vs. Stretch graph is equal to the spring constant. By finding the slope of each of your graphs, determine the spring constant k for each spring.

1. How do the graphs differ for the spring and the rubber band?

_____

_____

Going Further

Step 7: Repeat Steps 1 and 2 for two similar springs connected in series (end to end). Record the masses and strains in the last section of Table A. Determine the spring constant for the combination.

spring constant, k = _____

Table A

	SPRING 1			SPRING 2			SPRING 3		
MASS									
FORCE									
STRETCH									
SPRING CONSTANT									

	RUBBER BAND				SPRINGS IN SERIES			
MASS								
FORCE								
STRETCH								
SPRING CONSTANT								

Summing Up

2. How does the spring constant of two springs connected in a series compare with that of a single spring?

_____

_____

3. A child holds a slinky as shown. Why do the coils of the slinky get closer together towards the bottom?

_____

_____

By Hooke or By Crook

CONCEPTUAL **Physics** ▬▬▬▬▬▬▬▬▬▬▬▬▬▬▬▬▬▬▬▬▬▬▬ **Experiment**

Atomic Nature of Matter

Diameter of a BB

Purpose
To estimate the diameter of a BB.

Required Equipment and Supplies
75 mL of BB shot
100-mL graduated cylinder
tray
ruler
micrometer

Discussion
Eight wooden blocks arranged to form a 2 x 2 x 2 inch cube have an outer surface area that is less than the same eight blocks arranged in any other configuration. For example, if they are arranged to form a 1 x 2 x 4 inch rectangular block, the outer surface area will be greater. If the blocks are spread out to form a stack only one cube thick, 1 x 1 x 8 inches, the area of the surface will be greatest.

The different configurations have different surface areas, but the volume remains constant. The volume of pancake batter is also the same whether it is in the mixing bowl or spread out on a surface (except that on a hot griddle the volume increases because of the expanding gas bubbles that form as the batter cooks). The volume of a pancake equals the surface area of one flat side multiplied by the thickness. If both the volume and the surface area are known, then the thickness can be calculated from the following formulas:

$$\text{volume} = \text{area} \ \ \text{thickness}$$

$$\text{thickness} = \frac{\text{volume}}{\text{area}}$$

Instead of cubical blocks or pancake batter, consider a shoe box full of marbles. The space taken up by the marbles equals the volume of the box. Suppose you computed the volume and then poured the marbles onto a large tray. Can you think of a way to estimate the diameter (or thickness) of a single marble without measuring the marble itself? Would the same procedure work for smaller size balls such as BB's? Try it in this activity and see. It will be simply another step smaller to consider the size of molecules.

Procedure
Step 1: Use a graduated cylinder to measure the volume of the BB's. (Note that 1 mL = 1 cm^3.)

volume = _____ cm^3

Step 2: Carefully spread the BB's out to make a compact layer one pellet thick on the tray. Determine the area covered by the BB's. Describe your procedure and show your computations.

area = _____ cm^2

Step 3: Using the area and volume of the BB's, estimate the diameter of a BB. Show your computations.

estimated diameter = _____ cm

Step 4: Check your estimate by using a micrometer to measure the diameter of a BB.

measured diameter = _____ cm

Summing Up

1. What assumptions did you make when estimating the diameter of the BB?

_____

2. How does your estimate and the measurement of the diameter of the BB compare? Calculate the percentage difference (consult the Appendix on how to do this) between the measured and estimated diameter of the BB.

_____

3. Oleic acid is an organic substance that is soluble in alcohol but insoluble in water. When a drop of oleic acid is placed in water, it usually spreads out over the water surface to create a *monolayer*, a layer that is one molecule thick. Describe a method to estimate the size of an oleic acid molecule.

_____

_____

_____

_____

_____

Diameter of a BB

CONCEPTUAL Physics

Diameter of a Molecule

Experiment

Oleic Acid Pancake

Purpose
To estimate the size of a molecule of oleic acid.

Required Equipment and Supplies
tray
water
chalk dust or lycopodium powder
eye dropper
oleic acid solution (5 mL oleic acid in 995 mL ethanol)
10-mL graduated cylinder

Discussion
During this experiment you will estimate the *diameter* of a single molecule of oleic acid, to determine for yourself the extreme smallness of a molecule. The procedure for measuring the diameter of a molecule will be much the same as that of measuring the diameter of a BB in the previous activity. The diameter is calculated by dividing the volume of the drop of oleic acid used, by the area of the *monolayer* film that is formed. The diameter of the molecule is the depth of the monolayer.

volume = area depth

$$\text{depth} = \frac{\text{volume}}{\text{area}}$$

Step 1: Pour water into the tray to a depth of about 1 cm. Spread chalk dust or lycopodium power very lightly over the surface of the water; too much will hem in the oleic acid.

Step 2: Using the eye dropper, gently add a single drop of the oleic acid solution to the surface of the water. When the drop touches the water, the alcohol in it will dissolve in the water, but the oleic acid will not. The oleic acid spreads out to form a nearly circular patch on the water. Measure the diameter of the oleic acid patch in several places, and compute the average diameter of the circular patch. Then compute the area of the circle.

average diameter = _____ cm

area of circle = _____ cm^2

Step 3: Count the number of drops of solution needed to occupy 1 mL (or 1 cm^3) in the graduated cylinder. Do this three times, and find the average number of drops in 1 cm^3 of solution.

number of drops in 1 cm^3 = _____

Oleic Acid Pancake

Divide 1 cm^3 by the number of drops in 1 cm^3 to determine the volume of a single drop.

volume of single drop = _____ cm^3

Step 4: The volume of the oleic acid alone in the circular film is much less than the volume of a single drop of the solution. The concentration of oleic acid in the solution is 5 mL per liter of solution. Every cubic centimeter of the solution thus contains only $\frac{5}{1000}$ cm^3, or 0.005 cm^3, of oleic acid. The volume of oleic acid in one drop is thus 0.005 of the volume of one drop. Multiply the volume of a drop by 0.005 to find the volume of oleic acid in the drop. This is the volume of the layer of acid in the tray.

volume of oleic acid = _____ cm^3

Step 5: Estimate the diameter of an oleic acid molecule by dividing the volume of oleic acid by the area of the circle.

diameter = _____ cm

An oleic acid molecule is not spherical, but rather elongated like a hot dog. One end is attracted to water, and the other end points away from the water surface. The molecules stand up like people in a puddle! So the estimated diameter is actually the estimated length of the short side of an oleic acid molecule.

Summing Up
1. What is meant by a *monolayer*?

_____

2. Why is it necessary to dilute the oleic acid for this experiment?

_____

3. Which substance forms the monolayer film — the oleic acid or the alcohol?

_____

4. The shape of oleic acid molecules is more like that of a rectangular hot dog than a cube or marble. Furthermore, one end is attracted to water (or *hydrophyllic*) so that the molecule actually stands up on the surface of water. If each of these rectangular molecules is 10 times as long as it is wide, how would you estimate the volume of one oleic acid molecule?

_____

_____

Oleic Acid Pancake

CONCEPTUAL **Physics**   Experiment

Archimedes' Principle

Float a Boat

Purpose
To investigate Archimedes' principle and the principle of flotation.

Required Equipment and Supplies
spring scale
triple-beam balance
string
rock or hook mass
600-mL beaker
500-mL graduated cylinder
clear container or 3 gallon bucket
water
masking tape
chunk of wood
modeling clay
toy boat capable of a 1200 gram cargo, or 9 inch aluminum cake pan
100-g mass
3 lead masses or lead fishing weights

Discussion
An object submerged in water takes up space and pushes water out of the way.
We say the water is *displaced*. Interestingly enough, the water that is pushed
out of the way pushes back on the submerged object. For example, if the object
pushes a volume of water with a weight of 100 N out if its way, then the water
reacts by pushing back on the object with a force of 100 N — Newton's third
law. We say that the object is *buoyed* upward with a force of 100 N. This is
summed up in Archimedes' principle, which states that the *buoyant force* that
acts on any completely or partially submerged object *is equal to the weight of
the fluid the object displaces.*

Procedure
Step 1: Use a spring scale to determine the weight of an object (rock or hook mass)
that is first out of water and then under water. The difference in weights is the
buoyant force. Record the weights and the buoyant force.

weight of object out of water = _____

weight of object in water = _____

buoyant force on object = _____

Step 2: Devise a method to find the volume of water displaced by the object. Record
the volume of water displaced. Compute the mass and weight of this water.
(Remember, 1 mL of water has a mass of 1 g and a weight of 0.01 N.)

volume of water displaced = _____

mass of water displaced = _____

Float a Boat

weight of water displaced =_____

1. How does the buoyant force on the submerged object compare with the weight of the water displaced?

_____

_____

NOTE: To simplify calculations, for the remainder of this experiment measure and determine *masses*, without finding their equivalent *weights* $(W = mg)$. Keep in mind, however, that an object floats because of a buoyant *force*. This force is due to the *weight* of the water displaced.

Step 3: Measure the mass of a piece of wood with a beam balance, and record the mass in Table A. Measure the volume of water displaced when the wood floats. Record the volume and mass of water displaced in Table A.

2. What is the relation between the buoyant force on any floating object and the weight of the object?

_____

_____

_____

3. How does the mass of the wood compare to the mass of the water displaced?

_____

_____

4. How does the buoyant force on the wood compare to the weight of water displaced?

_____

_____

Step 4: Add a 100-g mass to the wood so that the wood displaces more water but still floats. Measure the volume of water displaced and calculate its mass, recording them in Table A.

5. How does the buoyant force on the wood and 100-g mass compare to the weight of water displaced?

_____

_____

Step 5: Roll the clay into a ball and find its mass. Measure the volume of water it displaces after it sinks to the bottom of a graduated cylinder. Calculate the mass of water displaced. Record all volumes and masses in Table A.

6. How does the mass of water displaced by the clay compare to the mass of the clay out of the water?

_____

_____

Float a Boat

7. Is the buoyant force on the submerged clay greater than, equal to, or less than its weight out of the water? What is your evidence?

_____

_____

Step 6: Retrieve the clay from the bottom, and mold it into a shape that allows it to float. Sketch or describe this shape. Measure the volume of water displaced by the floating clay. Calculate the mass of the water, and record in Table A.

Table A

OBJECT	MASS (g)	VOLUME OF WATER DISPLACED (ml)	MASS OF WATER DISPLACED (g)
WOOD			
WOOD AND 50-g MASS			
CLAY BALL			
FLOATING CLAY			

Summing Up

8. Does the clay displace more, less, or the same amount of water when it floats as it did when it sank?

_____

_____

9. Is the buoyant force on the floating clay greater than, equal to, or less than its weight when out of the water?

_____

_____

10. What can you conclude about the weight of an object and the weight of water displaced by the object when it floats?

_____

_____

11. Is the buoyant force on the clay ball greater when it is submerged near the bottom of the container or when it is submerged near the surface?

_____

_____

12. Is the pressure that the water exerts on the clay ball greater near the bottom of the container than when submerged near the surface?

_____

_____

13. Why are your last two answers different?

_____

Float a Boat

Going Further

Step 7: Suppose you are on a ship in a canal lock. If you throw a ton of bricks overboard from the ship into the canal lock, will the water level in the canal lock go up, down, or stay the same? Write down your prediction *before* you proceed to Step 8.

prediction for water level in canal lock: _____

_____

Step 8: Float a toy boat loaded with lead "cargo" in a relatively deep container filled with water (deeper than the height of the lead masses). For observable results, the size of the container should be just slightly bigger than the boat. Mark and label the water levels on masking tape placed on the container and on the sides of the boat. Remove the masses from the boat and put them in the water. Mark and label the new water levels.

14. What happens to the water level on the side of the boat when you place the cargo in the water?

_____

15. If a large freighter is riding high in the water, is it carrying a relatively light or heavy load of cargo?

_____

16. What happens to the water level in the container if you place the cargo in the water? Explain why this happens.

_____

_____

_____

17. Similarly, what happens to the water level in the canal lock when the bricks are thrown overboard?

_____

_____

18. Suppose the freighter is carrying a cargo of Styrofoam instead of bricks. What happens to the water level in the canal lock if the Styrofoam (which floats in water) were thrown overboard?

_____

_____

19. When a ship is launched at a shipyard, what happens to the sea level all over the world — no matter how imperceptibly?

_____

20. When a ship in the harbor launches a little rowboat from the dock, what happens to the sea level all over the world — no matter how imperceptibly?

_____

Float a Boat

CONCEPTUAL Physics

Thermal Expansion

Hot Stuff

Purpose
To investigate how a ball and ring expand when heated.

Required Equipment and Supplies
ball and ring expansion apparatus
bunsen or Fisher burner
600 mL beaker of water

Discussion
In this activity you will investigate the effects of heat on a solid metal ball and a metal ring. The ball just passes through the ring when both are the same temperature. Heat the metal ball in the flame of a bunsen burner and attempt to pass it through the ring. Does the ball pass through the ring? Why or why not?

_____

_____

_____

Procedure
Cool the ball by dipping it into in a beaker of water. This time heat the *ring*. Will the ball pass through the heated ring? Try it and see. Does the size of the hole increase, decrease, or remain the same?

_____

_____

_____

Summing Up
The ring expands in all directions when heated. Can you explain what happens and why?

_____

_____

_____

Hot Stuff

CONCEPTUAL **Physics**
Heat Transfer

Hot as a Griddle

Purpose
To debunk a popular myth.

Required Equipment and Supplies
ball and ring apparatus
Fisher (preferred) or bunsen burner
600 mL beaker

Discussion
You probably have heard about so-called firewalkers. Firewalkers are people who walk (or run!) on beds of hot coals, preferably with wet bare feet. How do they accomplish this apparent feat of mind over matter? Firewalkers are not harmed due to a combination of the low conductivity of ash and the short time allowed for the heat to conduct, plus the *Liedenfrost effect*.

If you never have tried it yourself, maybe you have sprinkled drops of water on a hot pancake griddle to test if it's ready. If the griddle isn't yet fully heated up, the water drops bubble and boil away. When the griddle is fully heated, the drops skitter about — seemingly floating above the griddle. This is an example of the Liedenfrost effect. The griddle is so hot that a layer of water vapor buffers the drop from the griddle, and to some degree, insulates it. The Leidenfrost effect accounts for the similar behavior of liquid nitrogen when it is poured onto a table top at room temperature.

Prcedure
Fill a 400 mL beaker with water. Warm the water until its about 90°C. Heat a metal ball until it's red-hot. **Caution: Do not touch anything or anybody with the ball!** While still red-hot, thrust the ball into the beaker of water. Try to hold it steady in the middle of the beaker. For best effect, you might try either turning down the room lights or turning them off completely. What do you observe?

_____

_____

_____

_____

Summing Up
How do you explain this phenomenon?

_____

_____

Hot as a Griddle

CONCEPTUAL Physics

Thermal Expansion

Hot Strip

Purpose
To observe expansion of a bimetallic strip.

Required Equipment and Supplies
bimetallic strip
bunsen burner
ice-water or liquid nitrogen (if available)

Discussion
When you dip a thermometer into a cup of hot coffee, the liquid mercury in the thin capillary tube rises. The reason the mercury rises in the thermometer is not that the glass does not expand — it does a little — but the mercury expands more than the glass. This is called differential expansion.

Procedure
Place a bimetallic strip into the flame of a Bunsen burner. What happens? Which side of the strip expands more?

_____

_____

Allow the strip to cool and it returns to its original shape. What will happen when the strip is dipped into ice water or liquid nitrogen?

_____

Try it and see. What happens?

_____

Summing Up

1. Suppose you are working at a machine shop and you wanted to fit a ring on a shaft so that it would not come off. If you heated the ring so that it just slipped on the cooled shaft, explain how you might go about getting the ring off again.

_____

_____

2. How are bimetallic strips employed to turn off devices such as electric motors, hairdryers, and toasters?

_____

Hot Strip

CONCEPTUAL **Physics** | **Activity**

Mechanical Equivalent of Heat

Niagara Falls

Purpose
To observe the effects of thermal agitation on temperature.

Required Equipment and Supplies
blender
thermometer

Discussion
When water flows over a waterfall, it loses PE and gains KE. As it crashes at the bottom of the falls, its KE is converted into heat. If all the PE of the water is converted into heat with no loss due to evaporation or by any other means, one kilogram of water that falls one meter increases its temperature by a little more than 0.002 °C. To see why, check the following calculation:

$$PE_{lost=}KE_{gained} = Heat_{gained}$$

$$PE_{lost} = Heat_{gained}$$

$$mgh(\text{in joules}) \sim mc\Delta T(\text{in calories})$$

If we express specific heat c in terms of joules, we can solve for ΔT. Then

$$mgh = mc\Delta T$$

$$\Delta T = \frac{gh}{c}$$

$$= \frac{(9.8\,\frac{m}{s^2}) \cdot (1m)}{4187\,\frac{J}{kg°C}}$$

$$= 0.0023°C$$

1. How much would the temperature rise if 2 kg fell one meter? 100 kg? 1 million kg? Show your calculations.

$\Delta T_2 =$ _____

$\Delta T_{100} =$ _____

$\Delta T_{1,000,000} =$ _____

2. Predict the temperature rise of water falling 50 meters over Niagara Falls? (Actually, the cooling effect of evaporation practically cancels this rise.) Show your calculations.

Procedure

Simulate Niagara Falls with minimum evaporation by pouring about 200 mL of water into a blender. Measure the temperature. Run the blender several minutes. Measure the temperature again.

T_{before} = _____ T_{after} = _____

Summing Up

3. Did it remain the same? Why or why not?

_____

_____

_____

_____

Niagara Falls

CONCEPTUAL Physics

Change of State

Old Faithful

Purpose
To observe a model of a geyser.

Required Equipment and Supplies
Pyrex® funnel
saucepan (a clear pyrex coffee percolator works great)
hot plate

Discussion
Have you ever wondered how a geyser like Old Faithful works? How come the water spurts out at more or less regular intervals? Why do geysers always spray steam? Interestingly enough, a geyser operates on the very same principles as a coffee percolator. This activity will allow you to investigate one of nature's many wonders.

Procedure
Place a pyrex funnel mouth down in a saucepan full of water so that the straight tube of the funnel sticks above the water. Position the rim of the funnel on a coin so water can get under it. Place the pan on a stove and watch the water as it begins to boil. Where do the bubbles form first? Why?

_____

_____

Summing Up
As the bubbles rise, they expand rapidly and push water ahead of them. The funnel confines the water, which is forced up the tube and driven out at the top. Now do you know how a geyser and a coffee perculator work? Explain their operation.

_____

_____

_____

_____

_____

Old Faithful

CONCEPTUAL Physics

Change of State

Boiling — A Cooling Process?

Purpose
To demonstrate that water can be boiled by lowering the pressure.

Required Equipment and Supplies
400 mL beaker
thermometer
vacuum pump with bell jar

Discussion
Have you ever noticed that is takes longer to cook by boiling in the mountains than it does down in the valley? Have you ever noticed that water boils in a *shorter* time when camping up in the mountains, but that it takes *longer* to cook potatoes in boiling water?

Procedure
Warm 200 mL of water in a 400 mL beaker to a temperature above 60°C. Record the temperature. Then place the beaker underneath the bell jar of a vacuum pump. If a thermometer will fit underneath the bell jar, place a thermometer in the beaker. Turn on the pump. What happens to the water?

T = _____

_____

_____

Was the water *really* boiling?

_____

_____

Stop the pump and remove the bell jar. What is the temperature of the water now?

T = _____

Summing Up

1. If boiling is *rapid* evaporation, is boiling a warming or cooling process?

_____

_____

2. Name two ways to cause water to boil.

_____

_____

3. How does perspiration aid a warm body?

_____

_____

CONCEPTUAL **Physics** | **Activity**

Change of State

Freezing — A Warming Process?

Purpose
To demonstrate an analog of the latent heat of fusion.

Required Equipment and Supplies
RE-HEATER® packs
hot plate
large pan or pot of boiling water

Discussion
We know that it is necessary to add heat to liquify a solid or vaporize a liquid. In the reverse process, heat is released when a gas condenses or freezes. Heat energy that accompanies these changes of state is called *latent heat of vaporization* (going from gas to liquid or liquid to gas), and *latent heat of fusion* (going from liquid to solid or solid to liquid).

Water, which usually freezes at 0°C or 32°F, can be found in a liquid state as low as –40°C (–40°F) or more. This *supercooled* water (liquid water below 0°C) often exists as cloud droplets that are very small. In fact, in order for snow or ice particles to form in a cloud, the temperature must be well below freezing because freezing depends on the presence of *ice-forming nuclei*. Ice-forming nuclei may be many different substances such as dust, bacteria, other ice particles, or silver iodide used to "seed" clouds during droughts. Most of these ice-forming nuclei are active in the range from –10°C to –20°C although silver iodide is active at temperatures as high as –4°C.

Cold clouds that contain large amounts of supercooled water and relatively small amounts of ice particles can be dangerous to aircraft. The skin of the aircraft, well below freezing, provides an excellent surface on which supercooled water can freeze. This is called aircraft icing. It can be quite severe under certain conditions.

The RE-HEATER provides a dramatic example of a supercooled liquid. What you observe in the RE-HEATER is actually the release of the latent heat of crystallization, which is analogous to the release of latent heat of vaporization or the latent heat of fusion. The freezing temperature of the sodium acetate solution inside the RE-HEATER is about 55°C (130°F), yet it exists at room temperature. The RE-HEATER can be cooled down to as low as –10°C before it finally freezes.

It only takes a quick click to activate the RE-HEATER. You will notice that the internal trigger button has two distinct sides. If you use your thumb and forefinger and squeeze quickly, you will not need to worry about which side is up.

After observing the crystallization of the sodium acetate and the heat released, you might want to try it again and measure the heat of crystallization. The package has a mass of about 146 grams. The packaging and the trigger mechanism have a mass of about 26 grams. Thus, the sodium acetate solution has a mass of about 120 grams.

Procedure

Place the RE-HEATER in an insulated container (such as a larger Styrofoam cup) with 400 g of room-temperature water. Allow the water and RE-HEATER to reach equilibrium. Measure and record its initial temperature. Activate the RE-HEATER button. After 10 or 15 seconds, maximum temperature will be obtained. Return the RE-HEATER to the container and measure the temperature of the water at regular intervals. How much heat is gained by the water?

Q = _____

Summing Up

1. How does the crystallization inside the RE-HEATER relate to heating and air conditioning in a building? *Hints:* Think about how steam heat and radiators work or how the refrigerants in an air conditioner work.

_____

_____

_____

2. How do these processes relate to airplane safety?

_____

_____

_____

3. What are some practical applications of the RE-HEATER?

_____

_____

_____

CONCEPTUAL Physics

Quantity of Heat

Specifically Water

Purpose
To predict the final temperature of a mixture of cups of water at different temperatures.

Required Equipment and Supplies
3 Styrofoam cups
liter container
thermometer (Celsius)
pail of cold water
pail of hot water

Discussion
If you mix a pail of cold water with a pail of hot water, the final temperature of the mixture will be between the two initial temperatures. What information would you need to predict the final temperature? This lab investigates factors that are involved in changes of temperature. First, your goal is to find out what happens when you mix *equal* masses of water at different temperatures.

Procedure
Step 1: Two pails filled with water are in your room, one with cold water and one with hot water. Fill one cup $\frac{3}{4}$ full with cold water from the pail. Mark the water level along the inside of the cup. Pour the cup's water into a second cup. Mark it as you did the first one. Pour the cup's water into a third cup, and mark it as before. Now all three cups have marks that show very nearly equal measures.

Step 2: Fill the first cup to the mark with hot water from the pail. Measure and record the temperature of both cups of water.

    temperature of cold water = _____

    temperature of hot water = _____

Step 3: Predict the temperature of the water when the two cups are combined. Then pour the two cups of water into the liter container, stir the mixture slightly, and record its temperature.

    predicted temperature = _____

    actual temperature of water = _____

1. If there was a difference between your prediction and your observation, what may have caused it?

Pour the mixture into the sink or waste pail. Do *not* pour it back into either of the pails of cold or hot water! Now, investigate what happens when *unequal* amounts of hot and cold water are combined.

Step 4: Fill one cup to its mark with cold water from the pail. Fill the other two cups to their marks with hot water from the pail. Measure and record their temperatures. Predict the temperature of the water when the three cups are combined. Then pour the three cups of water into the liter container, stir the mixture slightly, and record its temperature.

    predicted temperature = _____

    actual temperature of water = _____

Pour the mixture into the sink or waste pail. Do *not* pour it back into either of the pails of cold or hot water!

  2. How did your observation compare with your prediction?

    _____

  3. Which of the water samples (cold or hot) changed more when it became part of the mixture? Why do you think this happened?

    _____

    _____

Step 5: Fill two cups to their marks with cold water from the pail. Fill the third cup to its marks with hot water from the pail. Measure and record their temperatures. Predict the temperature of the water when the three cups are combined. Then pour the three cups of water into the liter container, stir the mixture slightly, and record the temperature.

    predicted temperature = _____

    actual temperature of water = _____

Pour the mixture into the sink or waste pail. Do *not* pour it back into either of the pails of cold or hot water!

  4. How did your observation compare with your prediction?

    _____

  5. Which of the water samples (cold or hot) changed more when it became part of the mixture? Why do you think this happened?

    _____

    _____

6. What determines whether the equilibrium temperature of a mixture of two amounts of water will be closer to the initially cooler or warmer water?

_____

_____

_____

7. The amount of internal energy in the water is proportional to what quantities?

_____

_____

_____

_____

CONCEPTUAL **Physics**

Activity

Specific Heat of Water

Spiked Water

Purpose
To predict the final temperature of water and nails when mixed.

Required Equipment and Supplies
Harvard trip balance
2 large insulated cups
bundle of short, stubby nails tied together with string
thermometer (Celsius)
hot and cold water
paper towels

Discussion
If you throw a hot rock into a pail of cool water, you know that the temperature of the rock will decrease. You also know that the temperature of the water will increase — but will its increase in temperature be more, less, or the same as the temperature decrease of the rock? Will the temperature of the water increase as much as the temperature of the rock goes down? Will the changes of temperature instead depend on how much rock and how much water are present?

Procedure
Step 1: Place a large cup on each pan of a beam balance. Place a bundle of nails into one of the cups. Add cold water to the other cup until it balances the cup of nails. When the two cups are balanced, the same mass is in each cup-a mass of nails in one, and an equal mass of water in the other. Visually determine if the amount of water in one cup is sufficient to submerge the nails in the other cup. If not, add more nails to the bundle so that the corresponding mass of water will completely submerge the nails.

Step 2: Set the cup of cold water on your work table. Remove the bundle of nails from its cup and place it beside the cup of cold water.

Step 3: Fill the empty cup 3/4 full with hot water. Lower the bundle of nails into the hot water. Be sure that the nails are completely submerged by the hot water. Allow the nails and the water to reach thermal equilibrium.

Step 4: Measure and record the temperature of the cold water and the temperature of the hot water around the nails.

1. Is the temperature of the hot water equal to the temperature of the nails? Why do you think it is or is not? Can you think of a better way to heat the nails to a known temperature?

_____

_____

_____

Spiked Water

Step 5: Predict what the temperature of the mixture will be when the hot nails are added to the cold water. Then lift the nails from the hot water and put them quickly into the cold water. Be sure that the nails are completely submerged by the cold water. When the temperature of the mixture stops rising, record it.

predicted temperature = _____

actual temperature = _____

2. How close is your prediction to the observed value?

_____

Step 6: Now repeat Steps 1 through 5 with hot water replacing cold water and cold nails. First dry the bundle of nails with a paper towel. Then balance a cup with the dry bundle of nails with a cup of *hot* water. Remove the nails and fill the cup 3/4 full with *cold* water. Record the temperature of the hot water in the first cup. Lift the nails from the cold water and *quickly* lower the bundle of nails into the hot water. Be sure that the nails are completely submerged. Predict what the temperature of the mixture will be when the cold nails are added to the hot water.

predicted temperature = _____

actual temperature = _____

Summing Up

3. How close is your prediction to the observed value?

_____

4. Discuss your observations with your partners and write an explanation for what happened.

_____

_____

_____

5. Suppose you have cold feet when you go to bed, and you want something to warm your feet throughout the night. Would you prefer to have a bottle filled with hot water, or one filled with an equal mass of nails at the same temperature as the water? Explain.

_____

_____

_____

6. Why does the climate of a mid-ocean island (such as Hawaii or even Iceland) stay nearly constant, with very little change in temperature throughout the year?

_____

_____

Spiked Water

CONCEPTUAL **Physics** | Experiment

Specific Heats of Substances

Specific Heats

Purpose
To measure the specific heats of common metals.

Required Equipment and Supplies
hot plate
specific heat specimens
beaker
tongs
Styrofoam cups
balance
thermometer

Discussion
Have you ever held a hot piece of pizza by its crust only to have the moister parts burn your mouth when you take a bite? The meats and cheese have high specific heat capacity, whereas the crust has a low specific heat capacity. How can you compare the specific heat capacities of different materials?

THE CRUST DOESN'T BURN ME BUT THE CHEEZE DOES...HMM...HOW COME?

In this experiment you will increase the temperature of metal specimens in boiling water and then place them in Styrofoam cups or calorimeters that contain a mass of water equal to the specimen at room temperature. The heat lost by the specimen equals the heat gained by the water.

$$Q_{lost} = Q_{gained}$$
$$m_s c_s \Delta T_s = m_w c_w \Delta T_w$$

The specific heat of the specimen is

$$c_s = \frac{m_w c_w \Delta T_w}{m_s \Delta T_s}$$

For water c_w = 1.00 cal/g·°C and if the mass of water is the same as that of the mass of the specimen, then the specific heat of the sample is

$$c_s = 1.00 \frac{cal}{g \cdot °C} \left(\frac{\Delta T_w}{\Delta T_s} \right)$$

Procedure
Step 1: Assemble as many pairs of Styrofoam cups with one cup inside the other as you have specimens. You have just constructed inexpensive double-walled calorie meters, called *calorimeters*. Measure the mass of each specimen. Since 1 mL of water has a mass of 1 g, carefully measure as many mL of tap water as there are grams for each specimen and place it in the calorimeter. Use a thermometer to measure the temperature of the water in the calorimeters.

Specific Heats

Step 2: Bring the water to a vigorous boil. Boil the specimens in the water more than a minute until you are convinced they are in thermal equilibrium with the water.

Step 3: Using tongs, quickly remove each specimen from the boiling water and place it in a calorimeter. Shake any droplets of water from the specimen. Record the final temperature of each specimen.

Summing Up

Compare your values for the specific heats of your specimens to those in Table A. How do your values compare?

_____

_____

Table A

SPECIMEN	MASS (g)	$T_{initial}$ (°C)	T_{final} (°C)	ΔT (°C)	C (cal/g °C)

Going Further

Try an unknown specimen and see how closely it matches the value of one in the table of specific heats.

Table A

SPECIFIC HEATS	
SUBSTANCE	$c\left(\frac{cal}{g\cdot°C}\right)$
ALUMINUM	0.215
COPPER	0.0923
GOLD	0.0301
LEAD	0.0305
SILVER	0.0558
TUNGSTEN	0.0321
ZINC	0.0925
MERCURY	0.033
WATER	1.000

THE SPECIFIC HEAT c = THE QUANTITY OF HEAT Q PER MASS m PER CHANGE IN TEMP. ΔT

$$c = \frac{Q}{m\Delta T}$$

THAT'S RIGHT FOR WATER.

$$c = \frac{1\,cal}{(1\,g\,°C)} = 1.0\,\frac{cal}{g\cdot°C}$$

Specific Heats

CONCEPTUAL **Physics**

Specific and Latent Heats

Cold Stuff

Purpose
To calculate the specific heat of a solid sample and the heat of vaporization of a liquid from the conservation of energy principle.

Required Equipment and Supplies
liquid nitrogen
balance scale, Harvard Trip preferred
50-g metal specimen
stopwatch
2 Styrofoam cups

Discussion
The law of energy conservation is far reaching. Where did it originate? It originated with experiments on heat performed in the 17th century by Count Rumford. This experiment gives the flavor of Rumford's work on heat using some really cold stuff!

Procedure
Part A: Determining the Specific Heat of a Specimen

Step 1: To determine the rate of evaporation of liquid nitrogen, pour about 180 - 200 grams of liquid nitrogen into two Styrofoam cups — one placed inside the other as shown. After measuring its precise mass using a balance scale, immediately start a stopwatch and time how long it takes 10 grams to evaporate. A convenient method of doing this is to set your balance for 10 grams *less* than the initial amount and watch for the balance arm to swing upwards. For example, if your initial amount of liquid nitrogen is 190 grams, set your balance to 180 grams and measure the time for the scale to balance again. Record the time it takes 10 grams of liquid nitrogen to evaporate.

    time = _____ g

Step 2: Calculate the rate of evaporation by dividing the mass by the time it took to evaporate.

    rate of evaporation = _____ g/s

Step 3: Immediately after 10 grams of liquid nitrogen have evaporated, place a 50-gram metal specimen in the LN (liquid nitrogen) and start the stopwatch. Assume the initial temperature of the metal mass is room temperature. After several minutes of rapid boiling, the LN will become "quiet" as it and the specimen reach thermal equilibrium. Remove the specimen and measure the mass of the remaining LN. Record your data below.

time that specimen was in the cup = _____ s

initial mass of LN, m = _____ g

final mass of LN, m = _____ g

total mass of LN that evaporated: m = _____ g

Step 4: Use the rate of evaporation calculated in Step 2 and the time measured in Step 3 to determine the mass of nitrogen that would have evaporated during the same amount of time had you *not* put the specimen in the cup.

mass due to evaporation = (rate of evaporation)·(time)

mass due to evaporation = _____ g

Step 5: The heat lost cooling the specimen equals the heat gained by the LN as it evaporates, the total mass of LN that evaporates while the specimen is in the cup is the mass of LN that cools the specimen plus the mass of LN that would have evaporated in the same time without having placed the metal specimen in the cups.

mass of LN that evaporates = (mass of LN to cool specimen) + (mass due to evaporation)

But the total mass of LN that evaporates while the specimen is in the cup is the difference in the mass of the cup.

(initial mass – final mass) = (mass of LN to cool specimen) + (mass due to evaporation)

Therefore, the mass of LN that cools the specimen is the difference in the mass of the cup less the amount that would have evaporated.

mass of LN to cool specimen = (initial mass – final mass) – (mass due to evaporation)

Calculate the mass of liquid nitrogen that evaporated to cool the metal specimen. Show your calculations.

mass of LN to cool specimen = _____ g

Step 6: The heat the specimen loses, whcih is the product of its mass, specific heat, and change in temperature $(m_s c \Delta t)$, equals the heat that causes the LN to evaporate $m_{LN} H_v$. Therefore,

$$m_s c \Delta T = m_{LN} H_v$$

If the heat of vaporization for LN, H_v = 47.5 cal/g, determine the specific heat of the metal specimen. Assume the specimen was originally at room temperature and was cooled down to the temperature LN boils, –196°C. Show your calculations.

$$c = \underline{\hspace{2cm}} \frac{cal}{g \cdot °C}$$

Part B: Determining the Heat of Vaporization of Nitrogen

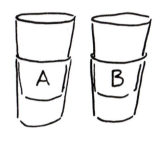

Step 1: Make two double-cup containers by placing one Styrofoam cup in side of another as shown. Label one combination "A" and the other "B". Carefully measure the mass of each container.

mass of container A, m = _____ g

mass of container B, m = _____ g

Step 2: Pour about 50-60 grams of warm (about 50°C) water in cup A. Measure and record its mass. Calculate the mass of the water by subtracting the mass of the cup you measured in Step 1. Then measure the initial temperature of the hot water and then remove the thermometer.

mass of water in A, m = _____ g

initial temperature of the water, T_i = _____ °C

Step 3: Pour about 40 grams of LN into cup B. Quickly (to minimize the mass loss due to evaporation) measure the mass of cup B (containing LN) and then pour the LN into cup A (containing warm water). A large white cloud of nitrogen vapor will be given off. Measure the time it takes to evaporate and calculate the mass of LN as you did for the water in Step 2.

mass of LN in B, m = _____ g

time to evaporate, t = _____ s

Step 4: When evaporation of the nitrogen is complete, gently stir the water with the thermometer until all the ice has melted. Measure and record the lowest temperature of the water.

final temperature of the water, T_f = _____ °C

Cold Stuff

Assuming the heat that cooled the water evaporated the LN, determine the heat of vaporization for nitrogen, H_v. Be sure your calculations take into account the amount the amount of LN that would have evaporated while the water was being cooled as you did in Step 2, Part A. Show your calculations.

$$H_v = \underline{\hspace{3cm}} \quad \frac{cal}{g}$$

Summing Up

1. How does your value for H_v compare with the accepted value of 47.5 cal/g?

_____

_____

CONCEPTUAL Physics ━━━━━━━━━━━━━━━━━━━━━━━━━━━━ **Experiment**

Inverse-Square Law

Solar Equality

Purpose

To measure the sun's power output and compare it with the power output of a 100-watt light bulb.

Required Equipment and Supplies

"Solar Equality" software or equivalent
2-cm by 6-cm piece of aluminum foil with one side thinly coated with flat black paint
clear tape
meterstick
two wood blocks
glass jar with a hole in the metal lid
one-hole stopper to fit the hole in the jar lid
thermometer (Celsius, range −10°C to 110°C)
glycerine
100-watt light bulb with receptacle

Discussion

If you told some friends that you had measured the power output of a light bulb, they probably would not be impressed. However, if you told them that you had computed the power output of the *sun* using only household equipment, they would probably be quite impressed — or maybe not even believe you. In this experiment, you will do just that — estimate the power output of the sun. You will need a sunny day to do this experiment.

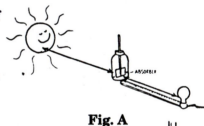

Procedure

Fig. A

Step 1: With the blackened side facing out, fold the middle of the foil strip around the thermometer bulb, as shown in Figure A. The ends of the metal strip should line up evenly.

Step 2: Crimp the foil so that it completely surrounds the bulb, as in Figure B. Bend each end of the foil strip outward (Figure C). Use a meterstick to make a flat, even surface. Use the unpainted side of a piece of clear tape to hold the foil on the thermometer.

METAL ENDS EVEN

Step 3: Use glycerin as a lubricant to insert the free end of the thermometer into the one hole stopper. Remove the lid from the jar, and place the stopper in the lid from the bottom side. Slide the thermometer until the foil strip is located in the middle of the jar. Place the lid on the jar.

Fig. B CRIMP HERE

Step 4: Position your apparatus indoors near a window so that the sun will shine on it. Prop it at an angle so that the blackened side of the foil strip is perpendicular to the rays of the sun. Keep the apparatus in this position in the midday sunlight until the maximum temperature is reached. If you prefer, position the apparatus outside. This, however, will necessitate that you do Steps 5 and 6 outside as well.

Fig. C

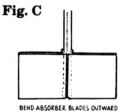

BEND ABSORBER BLADES OUTWARD

maximum temperature = _____ °C

Fig. D

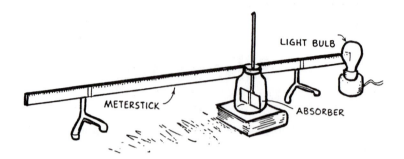

Step 5: Now you will find the conditions for bringing the apparatus to the same temperature with a clear 100-watt light bulb. Set the meter-stick on the table. Place a clear 100-Watt light bulb with its filament located at the 0-cm mark of the meter-stick (Figure D). Center the jar apparatus at the 95-cm mark with the blackened side of the foil strip perpendicular to the light rays from the bulb. You may need to put some books under the jar apparatus to align it properly.

Step 6: Turn the light bulb on. Slowly move the apparatus toward the light bulb, 5 cm at a time, allowing the thermometer temperature to stabilize each time. As the temperature approaches that was reached in Step 4, move the apparatus only 1 cm at a time. When the temperature obtained from the sun is maintained by the bulb for two minutes, turn the light bulb off.

Step 7: Measure as exactly as possible the distance in meters between the absorber strip of the apparatus and the filament of the bulb. Record this value.

distance from light filament to foil strip = _____ m

Step 8: For the distance obtained, use the following equation to compute the wattage of the sun. The Sun's distance in meters is 1.5×10^{11} m. Show your work.

$$\text{sun's wattage} = \frac{(\text{bulb's wattage})(\text{sun's distance})^2}{(\text{bulb's distance})^2}$$

sun's wattage = _____ W

Step 9: Use the value of the sun's wattage to compute the number of 100 watt light bulbs needed to equal the sun's power. Show your work.

number of 100 watt light bulbs = _____

Summing Up

1. Re-express the equation in Step 8 as a relationship between two ratios. Express the new equation as a sentence that begins, "The ratio of ... ."

_____

_____

2. The accepted value for the sun's wattage is 3.8×10^{26} W. List four factors that might account for the difference between your experimental value and the accepted value.

_____

_____

_____

_____

3. Would it be possible here on earth to turn on the number of 100-watt light bulbs you computed in Step 9 at once? Explain.

_____

_____

_____

Solar Equality

CONCEPTUAL **Physics**

Solar Energy

Experiment

Solar Power

Purpose
To find the daily amount of solar energy reaching the earth's surface and relate it to the daily amount of solar energy falling on an average house.

Required Equipment and Supplies
2 Styrofoam cups
graduated cylinder
water
blue and green food coloring
plastic wrap
rubber band
thermometer (Celsius)
meterstick

Discussion
How do we know how much total energy the sun emits each day? First we assume that, on the average, the sun emits energy equally in all directions. Imagine a heat detector so big that it completely surrounds the sun, like an enormous basketball. Then the amount of heat reaching the detector would be the same as the total solar output. Or if our detector were like half a basketball and caught half the sun's energy, then we would multiply the detector reading by 2 to compute the total solar output. If our detector encompassed a quarter of the sun and caught one fourth its energy, then its total output would be 4 times the detector reading.

Now that you have the idea, suppose that our detector is the surface area of a Styrofoam cup here on earth facing the sun. Then it comprises only a tiny fraction of the space that surrounds the sun. If you figure what that fraction is and also measure the amount of energy captured by your cup, you can tell how much total energy the sun emits. That's primarily how it's done! In this experiment, you will measure the amount of solar energy that reaches a Styrofoam cup and relate it to the amount of solar energy that falls on a housetop. You will need a sunny day for this.

Procedure
Step 1: Measure and record the amount of water needed to fill a Styrofoam cup.

volume of water = _____ mL

mass of water = _____ g

Nest the cup in a second Styrofoam cup. Fill it with the recorded amount of water. Add equal amounts of blue and green food coloring to the water until a dark liquid results. A dark liquid will be a good absorber of the sun's energy.

Step 2: Measure the water temperature and record it. Cover the cup with plastic wrap sealed with a rubber band.

     initial water temperature = _____ °C

Step 3: Put the cup in the sunlight for 10 minutes.

Step 4: Remove the plastic wrap. Stir the water in the cup gently with the thermometer, and record the final water temperature.

     final water temperature = _____ °C

Step 5: Find the difference in the temperature of the water before and after it was set in the sunlight.

     temperature difference = _____ °C

Step 6: Measure and record the diameter of the top of the cup in centimeters . Compute the surface area of the top of the cup in square centimeters.

     cup diameter = _____ cm

     surface area of water = _____ cm^2

Step 7: Compute the energy in calories that was collected in the cup. Assume that the specific heat of the mixture is the same as the specific heat of the water. Show your work.

     energy = _____ cal

THE QUANTITY G OF HEAT ENERGY COLLECTED BY THE WATER = MASS OF THE WATER × ITS SPECIFIC HEAT (c = 1 °/c) × ΔT, ITS CHANGE IN TEMPERATURE

Step 8: Compute the solar energy flux, the energy collected per square centimeter per minute. Show your work.

     solar energy flux = _____ $\frac{\text{cal}}{\text{m}^2} \cdot \text{min}^2$

Step 9: Compute how much solar energy reaches each square meter of the earth per minute. Show your work. (Hint: There are 10,000 cm^2 in 1 m^2.)

     solar energy flux = _____ $\frac{\text{cal}}{\text{m}^2} \cdot \text{min}^2$

Solar Power

Step 10: The distance between the earth and the sun is 1.5×10^{11}m, or 1.5×10^{13}cm.

The area of a sphere is $4\pi r^2$ so you can calculate the surface area of a "basketball" with radius 1.5×10^{13}cm. And you can divide this area by the surface area of the cup you used in this experiment to find what fraction of the sun's total solar output reached this cup. Then you can calculate the total solar output per minute from your work in Step 8. Do so.

    total solar output per minute = _____ cal

Step 11: Compute how much solar energy would fall in one 8 hour sunny day on a 4 m by 15 m roof that faces south. Show your work.

    energy received by roof = _____ cal

Summing Up

Scientists have measured the amount of solar energy flux just above our atmosphere to be 2 calories per square centimeter per minute (equivalently 1.4 kW/m^2). This energy flux is called the solar constant. Only 1.5 calories per square centimeter per minute reaches the earth's surface after passing through the atmosphere.

1. What factors could affect the amount of sunlight reaching the earth's surface and decrease the solar constant?

_____

_____

_____

_____

_____

_____

_____

_____

_____

_____

_____

Solar Power

CONCEPTUAL **Physics**

Vibration and Waves

Tuning Forks Revealed

Purpose
To observe and explore the oscillation of a tuning fork.

Required Equipment and Supplies
variety of tuning forks (low frequency (40-150 Hz) forks work best for large amplitude)
strobe light, variable frequency

Discussion
The tines of a tuning fork oscillate at a very precise frequency. That's why muscians use them to tune instruments. In this activity you will investigate their motion with a special illumination system — a stroboscope.

Procedure
Strike a tuning fork with a mallet or the heel of your shoe (do NOT strike against the table or other hard object). Does it appear to move?

_____

_____

Repeat, but this time immerse the tip of the tines just below the surface of water in a beaker. What do you observe?

_____

_____

Now dim the room lights and strike a tuning fork while it is illuminated with a strobe light. For best effect, use the tuning fork with the longest tines available. Adjust the frequency of the strobe so that the tines of the tuning fork appear to be stationary. Then carefully adjust the strobe so that the tines slowly wag back and forth. Describe your observations.

_____

_____

_____

Summing Up

What happens to the air next to the tines as they oscillate?

_____

_____

Strike a tuning fork and observe how long it vibrates. Repeat placing the handle against the table top or counter. Although the sound is louder, does the *time* the fork vibrates increase or decrease? Explain?

_____

_____

Imagine you struck the tuning fork in outer space — what would happen then?

_____

_____

CONCEPTUAL Physics

Vibration and Waves

Sound Off

Purpose
To demonstrate that sound does not travel in a vacuum.

Required Equipment and Supplies
1500 mL flask
one-hole rubber stopper
wire
Christmas-type bell
piece of paper
matches

Discussion
At the movies we see the huge space ship roar across the screen in its interplanetary journey. Does it really roar?

Procedure
Obtain a 1000 - 2000 mL flask. Suspend a small Christmas-type bell by a piece of wire attached to a rubber stopper so the bell swings freely when inside the flask. Make sure the hole in the stopper is corked with a small amount of masking tape wrapped around the wire. Shake the flask and listen to the sound of the ringing bell. Note its loudness and pitch. Wad up a small piece of paper in the shape of a cigar. Light one end with a match then quickly thrust the burning paper in the flask. Allow the paper to burn. As the flame subsides, place the bell in the flask and give the stopper a slight twist so that it forms a good seal. Now shake the flask to ring the bell.

RUBBER STOPPER

CHRISTMAS BELL

Can you hear the bell ring now? Why or why not?

_____

_____

_____

What change do you "note" in the sound of the bell?

_____

_____

_____

Sound Off

CONCEPTUAL Physics | **Activity**

Speed of Sound in Air

Sir Speedy

Purpose
To estimate the speed of sound using an echo.

Required Equipment and Supplies
large building with a flat wall
stopwatch or wrist watch with second hand

Discussion
This activity is similar to one Robert Millikan — who was the first American physicist to win the Nobel prize — did with his students at CalTech. Do and sound off!

Procedure
If you can find a flat side of a building with a good echo, have your partner clap their hands until you can hear successive echoes clearly. Position yourself so that you can measure the time from the clap to the last echo you can distinctly hear.

One method of doing this is to clap steadily, adjusting your rate until each reflection is heard exactly midway between the preceding and following claps. Once you get the rhythm, with a stopwatch measure the time it takes to clap 10 to 20 times.

The distance traveled during that time will be the number of echoes multiplied by the round-trip distance to the wall or end of the corridor. Estimate the speed of sound by dividing this distance by the time.

$v = $ _____

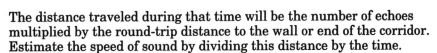

Summing Up
How close are you to the accepted value of 340 m/s at 20°C?

_____

_____

CONCEPTUAL **Physics**

Standing Waves in a Tube

Give Sound a Whirl

Purpose
To observe the partial-tones in a hoot tube.

Required Equipment and Supplies
hoot tubes (Whirly®) of various lengths
smooth tube (garden hose works fine)
signal generator, audio amplifier with speaker

Discussion
Sound is the vibrations of air. One way to make air vibrate is to send
air currents along the corrugations of a flexible tube, or "hoot" tube.
Have fun hooting!

Procedure
Whirl a hoot tube and generate as many different tones as you can.
How is the speed of the tube related to the pitch? Try a hose with no
corrugations. Does it hoot? Why do you suppose air is made to vibrate
in one tube and not the other? Unlike the standing waves in a wide-
mouth tube that have a node at the closed end and an antinode at the
open end both ends of a hoot tube are open and are therefore antinodes.
The standing waves for the first two harmonics are illustrated below.
To better understand why different tones resonate in the tube, sketch
standing waves that represent the third and fourth harmonics in a
similar fashion.

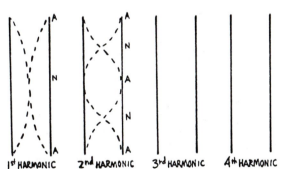

1st HARMONIC   2nd HARMONIC   3rd HARMONIC   4th HARMONIC

Use masking tape to close one end and try again. Does the tube hoot?
Why or why not?

_____

_____

Hoot Tube Harmonics

For a tube open at both ends, the standing waves that can be produced are the fundamental and its harmonics — multiples of the fundamental (the first harmonic f_1). That is, $f_n = nf_1$. For example, if the fundamental frequency, f_1, is 100 Hz, the successive harmonics are 200, 300, 400, 500, 600, . . .1000 Hz, etc. The harmonic sequence has a musical pattern like C(1), C'(2)[*], G(3), C"(4), E(5), G(6), Bb(7), C'''(8), D(9), E(10). After you've had a chance to practice your whirling technique, compare your generated whirly tones to those created by a signal generator connected to an audio amplifier. The signal generator will give you the frequency in hertz. The first harmonic is difficult to hear — it's barely audible. To hear it, you will need to whirl the tube with the stationary end snug against your ear (this may be a little awkward to do!) and match it to the tone generated by a signal generator. Successive harmonics are much louder. The second harmonic is often mistaken for the first.

Match the successive harmonics of the hoot tube in a similar manner to the tones of the signal generator. Record your results and complete the blank space in Table A.

Table A

HARMONIC	PREDICTED HARMONIC FREQUENCY $f = \frac{v}{\lambda} = \frac{340}{2L} = f_o$	TONE OF SIGNAL GENERATOR (Hz)
1st	$= \frac{340}{L} = (2)\frac{340}{2L} = 2f_o$	
2nd	$= \frac{340\frac{1}{2}}{3L} = (3)\frac{340}{2L} = 3f_o$	
3rd	$= \frac{340\frac{1}{L}}{2} = (4)\frac{340}{2L} =$	
4th	$=$	

Summing Up

How well do the predicted frequencies of the hoot tube harmonics compare with the matched tones of the signal generator?

_____

_____

_____

[*] C'= the C one octave above C; C" = the C two octaves above C, etc.

CONCEPTUAL Physics

Waveforms of Sound

Oh Say Can You Sing?

Purpose

To observe what different sounds look like on an oscilloscope.

Required Equipment and Supplies

oscilloscope
microphone (depending upon the microphone, a small pre-amp may be necessary)
as many different musical instruments as possible

Discussion

We know the source of sound is a vibration, which we normally
hear but do not see. Now we'll experience both.

Procedure

Make various sounds for the microphone that has been
connected to an oscilloscope. Your instructor can fill you in
on the operation of an oscilloscope. Try singing the vowels,
whistling, clapping, etc. What happens to the shape of the
waves on the oscilloscope screen as you increase the pitch?

_____

_____

How does the shape of the waves made by your singing compare to those of a tuning
fork? Clapping?

_____

_____

Sketch several of these different waveforms.

Now play notes from various musical instruments that are available to you, such as a recorder, cello, violin, drum, guitar, flute, etc. What do they look like?

Summing Up

Can you see any common pattern among the different sounds?

_____

_____

_____

_____

CONCEPTUAL **Physics**   █ **Experiment** █

Resonance of an Air Column

Sound Barrier

Purpose
To determine the speed of sound using the concept of resonance.

Required Equipment and Supplies
golf-club tube cut in half
1-Liter graduated cylinder
meterstick
tuning forks between 256 & 400 Hz
Alka-Seltzer® tablet (optional)

Discussion
You are familiar with many applications of resonance. You may have
heard a vase on a shelf across the room rattle when a particular note on
a piano was played. The frequency of that note was the same as the
natural vibration frequency of the shelf. Your textbook gives other
examples of resonating solid objects.

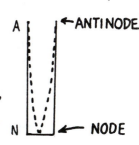

Gases also resonate; standing waves of air can be induced in organ
pipes, flutes, and soda-pop bottles. A vibrating tuning fork held over an
open tube may vibrate the air column in it at its resonant frequency. The length
of the air column can be shortened by adding water to the tube. The volume of the
sound becomes loudest at the proper length for resonance. The frequency of the
vibrating air column and the vibrating tuning fork match. In general, for a tube open
at one end and closed at the other, acoustical resonance occurs when the air column is
one-fourth the wavelength of the sound wave.

In this experiment you will use the concept of resonance to determine the wavelength
of a sound wave of known frequency. You can then compute the speed of sound, since
the speed of any wave equals the product of the frequency and wavelength, $v = f\lambda$.

Displacement vs. Pressure
Drawing sketches of standing waves helps us to visualize how resonance occurs.
Most of the time we refer to the motion or *displacement* of the molecules. For a
tube open at one end and closed on the other, the air molecules at the open end
of the tube are able to jiggle back and forth, whereas the air molecules next to
the closed end can't. Thus, the open end of the air column is a displacement
antinode and the closed end a displacement *node*.

Helpful as these displacement diagrams are to visualize the motion of molecules,
they do not help us understand why certain notes resonate while others do not.
To understand better how acoustical resonance comes about, imagine a pressure
pulse traveling down a tube with an open end generated by vibrating lips or
reeds at the closed end of the tube.

First, a high pressure pulse is generated by the player and forces the reed to close as
the pulse continues down the tube. The high pressure pulse reaches the open end (a
total distance *L* so far) and is reflected just as a pulse traveling down a slinky does
when it reaches its end. The reflected pulse, however, changes phase 180° and becomes
a low pressure pulse — too low to push the reed open when it arrives at the closed end
of the tube (total distance traveled 2*L).*

Sound Barrier

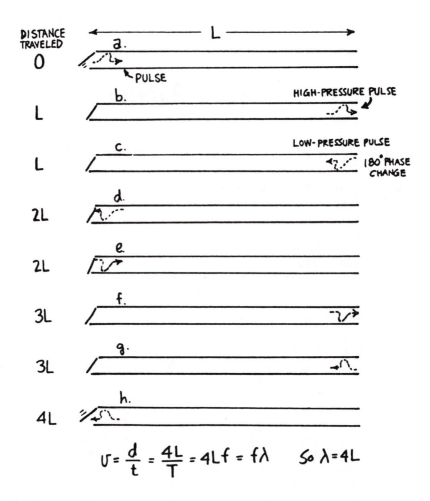

$$v = \frac{d}{t} = \frac{4L}{T} = 4Lf = f\lambda \qquad \text{So } \lambda = 4L$$

At the closed end, the pulse reflects in phase, and continues toward the open end as a low pressure pulse (total distance traveled now $3L$). A phase change upon reflecting from the open end converts it to a high pressure pulse. This high pressure pulse travels back to the reed ($4L$), pushes on it with sufficient pressure to swing it open again. That's why the pulse makes two round trips for a total distance of $4L$ each time the lip or reed vibrates at frequency f once during a time interval of one period, T. The speed ($v = distance/time$) of the wave is: $v = 4L/T = 4Lf$. The length of the tube, therefore, is $\lambda/4$.

When you tighten your lips and blow on a trumpet or other wind instrument, your lips are capable of vibrating at all sorts of frequencies. The air column, however, "selects" certain frequencies and resonates. The vibration of your lips (or reed in the case of clarinets, etc.) is regulated by the "acoustic feedback" of the air column itself. That is, the air column helps your lips vibrate at a particular frequency. . . your lips provide energy to vibrate the air column . . .which in turn helps your lips vibrate. . . voila — *resonance!*

Displacement is not to be confused with pressure. A displacement *antinode* is at a pressure *node* and vice-versa. The molecules near the closed end of the tube are compressed by molecular bombardment whereas the molecules near the open end remain at the same pressure.

Sound Barrier

Procedure

Step 1: Fill the cylinder with water to two thirds of its capacity. Place the resonance tube in the cylinder. You can vary the length of the air column in the tube by moving the tube up or down.

Step 2: Select a tuning fork, and record its frequency.

frequency = _____ Hz

Strike the tuning fork with a rubber mallet or the heel of your shoe (NOT on the cylinder). Hold the tuning fork horizontally, with its tines one above the other, about 1 cm above the open end of the tube. Move both the fork and the tube up and down together to find the air column length that gives the loudest sound. There are several loud spots, but you are to locate the very loudest one with the shortest open tube length. Mark the water level in the tube for this loudest sound.

Step 3: Measure the distance from the top of the resonance tube to the water level marked.

length of air column = _____ m

Step 4: Measure the diameter of the resonance tube.

diameter of the resonance tube = _____ m

Step 5: Make a corrected length by adding 0.4 times the diameter of the tube to the measured length of the air column. This corrected length accounts for the small amount of air just above the tube that also vibrates.

corrected length = _____ m

Step 6: The corrected length is one fourth of the wavelength of the sound vibrating in the air column. Compute the wavelength of that sound.

wavelength = _____ m

Step 7: Using the frequency and the wavelength of the sound, compute the speed of sound in air. Show your work.

speed of sound in air = _____ m/s

Step 8: If time permits, repeat Steps 2 to 7 using a tuning fork of different frequency.

frequency = _____ Hz

length of air column = _____ m

corrected length = _____ m

wavelength = _____ m

speed of sound in air = _____ m/s

Sound Barrier

Summing Up

1. The accepted value for the speed of sound in air is 332 m/s at 0°C. The speed of sound in air increases by 0.6 m/s for each degree Celsius above zero. Compute the accepted value for the speed of sound at the temperature of your room.

$v =$ _____ m/s

2. How does your computation for the speed of sound compare with the accepted value? Compute the percentage error (consult the Appendix on how to do this if necessary).

percentage error = _____

Going Further

When alka-seltzer is mixed with water, it chemically reacts with water and releases carbon dioxide gas (CO_2). Add an alka seltzer tablet to the water in the cylinder. The carbon dioxide gas will displace the air inside the column. Why?

_____

While the tablet is fizzing, predict whether the length of the new column will be longer, shorter, or the same length as before. When the fizzing stops, repeat the experiment. What is the speed of sound in CO_2?

predicted speed, $v =$ _____ m/s

actual speed, $v =$ _____ m/s

Sound Barrier

CONCEPTUAL **Physics** ─── Experiment

Resonance of an Aluminum Rod

Screech!

Purpose
To determine the speed of sound using the concept of resonance.

Required Equipment and Supplies
aluminum rod (about a meter long and 13 mm in diameter)
rosin
signal generator and audio amplifier

Discussion
Bells, chimes, and even tuning forks ring. But how about a
ordinary aluminum rod? Try it and see!

Procedure
Step 1: Hold an aluminum rod (about a meter long and 10-15 mm in diameter) with
your thumb and index finger in the middle of the rod. Liberally apply rosin to the
fingers of your other hand and slowly stroke the rod. Adjust the pressure of your
fingers until the rod begins to squeal. As the rod begins to build up amplitude, the
sound becomes quite impressive!

What vibration is causing the sound? Since you are clasping it at the middle, that
point must be a displacement *node*. What about each end?

Step 2: Sketch a standing wave which represents the displacement of the aluminum
atoms in the rod (antinodes at the ends; node in the middle).

1. What fraction of a wavelength is the length of the bar?

_____

_____

$\lambda =$ _____

Step 3: Measure and record the length of the bar.

length of the bar, $l =$ _____

Step 4: Calculate the wavelength.

wavelength, $\lambda =$ _____

Step 5: Now you are ready to calculate the corresponding frequency. Recall the wave
equation

$$v = f\lambda$$

Screech!

If f_1 represents the frequency of the first harmonic and the speed of sound in aluminum is about 5000 m/s, then

$$f_1 \cdot \lambda = 5000$$
$$f_1 \cdot (2L) = 5000$$
$$f_1 = \frac{5000}{2L}$$

$f_1 =$ _____ Hz

Step 6: Stroke the bar and make it resonate. Have your lab partner adjust the frequency knob of a signal generator (set up by your instructor) so that the two tones match. This may take some practice.

Summing Up

2. How does your predicted frequency compare to the one emitted by the signal generator?

_____

_____

3. Try adjusting the frequency of the signal generator so that the frequency is just a little higher or a little lower than the resonant frequency of the bar. Can you hear the beats (warbling sound)?

_____

_____

CONCEPTUAL **Physics**

Experiment

Partial Tones

Frequency Analyzer

Purpose
To observe partial-tones of different instruments.

Required Equipment and Supplies
Good Stuff software or equivalent
computer with analog to digital converter (A/D–D/A board)
microphone (depending on the microphone, a small pre-amp may be necessary)

Discussion
What is it that makes a flute sound different from a violin? Or the word "boot" sound different from "bat"? To see the partial tones that give sounds their character, your instructor has set up a computer with an analog to digital converter, a microphone, and an audio amplifier. Your instructor can explain the operation of this device for you if you are interested.

Procedure
Sing or play a steady note into the microphone. Adjust the pitch to obtain the clearest display on the computer. The waveform is shown near the top of the screen and the frequencies of the partial tones present are graphed below.

What does the partial tone series look like for the word "boot"? "Bat"? Make a sketch of the waveforms.

_____

_____

How does it look for a tuning fork?

_____

_____

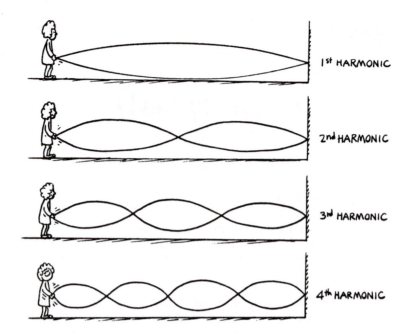

1st HARMONIC

2nd HARMONIC

3rd HARMONIC

4th HARMONIC

Summing Up

Play the same note on various instruments into the microphone. How do their partial tones series compare?

_____

_____

_____

_____

Does a violin have more partial tones than a flute or recorder?

_____

_____

_____

_____

CONCEPTUAL Physics ━━━━━━━━━━━━━━━━━━━━━━━━━ | **Activity**
Electric Charge

Charge It, Please!

Purpose
To observe the effects and behavior of static electricity.

Required Equipment and Supplies
Van de Graaff generator
several new balloons
matches
Styrofoam lunch tray
Styrofoam cup
cat's fur
Styrofoam peanuts or fresh puffed rice
light-weight aluminum pie tin

Discussion
You've probably walked across a carpet or pivoted out of a carseat and
been shocked when you reached for the door knob many times. Did you
know that the electrical charge that makes up the spark can be several
thousand volts? That's why technicians have to be so careful when
working around computer chips!

Procedure
Step 1: Stand on an isolation stand (or rubber mat) next to a
discharged Van de Graaff generator. Place one hand on the conducting
sphere on top of the generator and have your partner turn on the
generator motor on. Shake your head as the generator charges up.
What do you feel? Does your hair stand on end? Can you explain this
peculiar behavior?

_____

_____

_____

_____

Step 2: Discharge the generator by touching it with your finger or knuckle. Place a
small cup of puffed rice on top of the conducting sphere. Turn on the generator. What
happens? Why?

_____

_____

_____

_____

Charge It, Please!

Step 3: Light a wooden match and move it near a charged sphere on top of the generator. What do you observe? Explain.

_____

_____

Step 4: Blow-up a balloon. Vigorously stroke it against your hair. Can you pick up a Styrofoam peanut or some puffed rice with it? How many? Sketch how the charges might be arranged on the balloon to the right.

_____

_____

_____

Step 5: Vigorously stroke your hair with the balloon again. Can you cause the balloon to stick to the wall? Where and where not? What causes it to stick? Show how the charge might be situated on the balloon and the wall.

_____

_____

_____

_____

Step 6: Stroke one side of the balloon against your hair. Place the opposite side of the balloon against a wall where it stuck before. Let the balloon go. What do you observe? Is the charge on the balloons of the same sign or different sign?

_____

_____

_____

Step 7: Blow up a second balloon. Rub both balloons against your hair. Do they attract or repel each other?

_____

_____

_____

_____

Step 8: Tape a Styrofoam cup to a pie tin. Using masking tape, secure a Styrofoam lunch tray to your table. Vigorously rub cat's fur onto the Styrofoam lunch tray. Rub a balloon on your hair and bring it close to the tray. Is it attracted or repelled by the tray?

_____

Step 9: Charge the lunch tray with the cat's fur. Using the cup as a handle, place the pie tin on the lunch tray. Touch your finger to the pie tin. What do you observe?

_____

Step 10: Lift the pie tin off the lunch tray. Is the pie tin charged? To check, rub a balloon on your hair and see if it is attracted to or repelled by the pie tin.

_____

_____

Step 11: Discharge the pie tin with your finger. Place the pie tin back on the tray and repeat. Is charge conducted off the pie tin when you touch it with your finger? Is the pie tin charged as before?

_____

_____

Summing Up

1. Is charge being conducted from the tray?

_____

_____

2. Can you explain how this happens?

_____

_____

CONCEPTUAL Physics

| | Activity |

Electrostatics

Sticky Electrostatics

Purpose
To investigate the nature of static electricity.

Required Equipment and Supplies
tape (Scotch brand Magic™ tape works well)
fresh balloons
digital voltmeter (DVM), with 10 megaohm input impedance or greater

Discussion
When discussing static electricity, many people focus on the need to rub materials together in order to generate separations of charge. Some state that friction creates the separation of charge. Is this statement true — or only some of the time? What *is* the nature of electrostatic charge?

Forget rubbing for now and consider two objects simply touched together. Their surfaces adhere slightly. Chemical bonds form at the regions of contact between the molecules of both surfaces. If the surfaces are not of the same material, the bonds will probably be *polarized*, with the shared bonding electrons staying with one surface more than with the other. When the two objects are pulled apart again, the bonds rupture and one surface may end up with electrons from the other surface. Now the surfaces are no longer neutral. One surface has extra electrons (electron *surplus*) while the other surface has fewer electrons (electron *deficient*) compared to the number of protons in each substance. The unbalanced charges are then separated as the surfaces are separated. The charges that formed electrically neutral molecules have been sorted into two groups and pulled apart by a great distance.

If the surfaces involved are rough or fibrous, friction does play a part in surface charging. If you touch a balloon to a head of hair, the hair really only touches the balloon in tiny spots, and the total area of contact is extremely small. However, if the balloon is *dragged* across the hair, then the successive areas of contact add up. Rubbing a balloon on your head increases the total area of contact, so it increases the amount of charge that is separated. However, the friction does not *cause* the charging. You can rub two balloons together as much as you like, and you will never create any "static electricity." Contact between *dissimilar* materials is required.

Procedure
Step 1: Pull a couple of strips of plastic adhesive tape from a roll. Each one should be about 12-20 cm long. Hold them up by their ends, then slowly bring them side by side. What happens? Notice that they repel each other. If you try to get the dangling lengths of tape to touch each other, the tape will swerve and gyrate to frustrate your efforts. Obviously the tape has become electrically charged. But how? No friction was involved.

Step 2: One at a time, pass each of the strips of tape lightly between your fingers, then hold the two strips near each other again. Now how do the strips behave? This time they will not repel each other. You've managed to discharge the strips by touching them.

Step 3: Fold over a couple of centimeters of the end of each strip. This gives you a non-sticky handle to work with. Carefully stick the two strips to each other so the sticky side of one strip adheres to the "dry" side of the other. You should end up with a double-thick layer of tape which is sticky on one side, and which has two tabs at the end. Now grasp the tabs and rapidly peel the strips apart. Keep them distantly separated, then slowly bring them together again. How do the strips behave this time?

Step 4: Blow-up two fresh balloons. Do not rub them against your hair or clothing. See if you can "create" static electricity by rubbing two electrically neutral balloons together. What is your result?

Step 5: Discharge your two strips of tape by running each one between your fingers. Hold them near each other to verify that they neither attract nor repel one another. Now stick the two strips together, but this time do it with the adhesive sides facing one another. Peel the strips apart again, then bring them near each other. Are the strips now neutral, or do they attract one another?

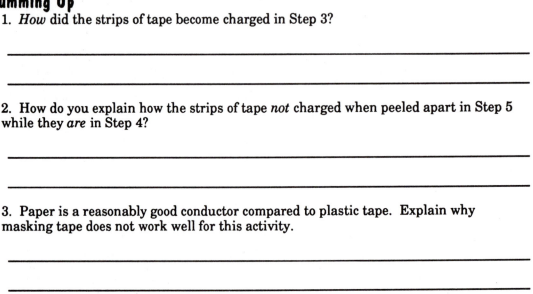

Step 6: Finally, peel four separate strips from a roll and neutralize them with your fingers. Then stick them together in pairs with the sticky side of one stuck to the dry side of the other. Now peel each pair apart so you have four charged strips.

Hold pairs of these strips together in different combinations. What do you discover? You can determine the polarity of the four strips by rubbing a balloon on your hair (rubber always acquires a negative charge when touched to hair) then holding it near the strip being investigated. If the balloon and the tape strip repel, the strip is negatively charged.

Summing Up

1. *How* did the strips of tape become charged in Step 3?

_____

_____

2. How do you explain how the strips of tape *not* charged when peeled apart in Step 5 while they *are* in Step 4?

_____

_____

3. Paper is a reasonably good conductor compared to plastic tape. Explain why masking tape does not work well for this activity.

_____

_____

Sticky Electrostatics

Going Further

A digital voltmeter (DVM) can be used to detect the sign of the charge that accumulates on an object. Turn the "DC Volts" scale of the meter to maximum sensitivity (some newer models are auto-ranging and only require being set to "DC Volts". When brought near a charged object, the digital display not only indicates a small voltage, but the *sign* of the charge on the object as well. Place the common or ground probe off to one side out of the way. The meter may give a small positive or negative readout — depending on its environment. Allow the meter to stabilize. Use the positive probe to investigate the sign of the charge on objects. When a positively charged object is brought near the probe, the positive charge on the object "pushes" on positive charges on the probe through the meter, causing the reading on the meter to *increase*. When a negatively charged object is brought near the probe, the negative charge on the object "pushes" on negative charges on the probe, causing the reading on the meter to *decrease*. Therefore, the sign of the charge on the object is related to the *change* of the reading on the meter—regardless of the sign of the original readout. As you investigate the sign of the charge of objects, you will note that the change in the readout "decays". The reason the readout of the meter decays is due to the transient nature of how the meter works—not because the the charge on the object is dissipating. The decay rate depends on the input impedance of your particular meter. The greater the input impedance, the slower the decay. The lower the input impedance, the faster the decay.

Pull a 12-20 cm strip of tape from a roll. Being careful not to neutralize the tape, use a DVM to determine the sign of the charge on the strip of tape. Repeat for several strips. Record your results below.

sign of the charge on the strip _____

Use the DVM to probe the roll of tape after pulling a strip of tape from the roll. What do you find?

CONCEPTUAL Physics

Electric Circuits

The Electric Ferry

Purpose
To explain that electric charges can move from one place to another, and classify some common materials as conductors or insulators.

Required Equipment and Supplies
3 one pound clean coffee cans or equivalent
metal thumbtack
piece of thin thread
bit of masking tape
2 paraffin blocks
piece of wood
bare wire
aluminum pie tin
small fluorescent tube (optional)

Discussion
Static electricity is often defined to be electric charge that is not "flowing". But in the case of lightning large amounts charge flow in a very short time. In this activity, you will be able to observe electric charge ferried from one location to another.

Fig. A

Procedure
Step 1: Set a metal can on the block of paraffin wax. Stand another can about 2 cm away from the first can. Place wood (or cardboard) on top of either one of the cans. Do not let it touch the other can. Fasten one end of same thread to the point of a thumbtack. Use a bit of masking tape for this. See Figure A. Tape the thread to the top of the wood. Have the thumbtack hang between the two cans about half way down as shown. The hanging tack is an electric "ferry," and should not touch either can.

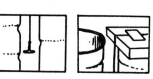

Now vigorously rub a Styrofoam lunch tray with cat's fur. Using a Styrofoam cup as a handle, place a pie tin on the lunch tray. Touch your finger to the pie tin. Touch the charged pie tin to the can on the wax. Repeat until the "ferry" begins to swing back and forth. Write down your observations.

_____

_____

_____

The Electric Ferry

Step 2: Set up a third can on a wax stand less than a meter away **Fig. B** as in Figure B. Connect the cans on the wax stands with bare copper wire. Let the wire hang between these cans and not touch anything else. Now touch the third can with a charged pie tin. Does the ferry go? How does it work?

_____

_____

Step 3: Again charge the ferry on the third can set on wax. What happens when the wire is removed?

_____

_____

_____

Step 4: This time connect the cans on wax with materials other than wire. Try long pieces of string, wood, metal, and glass. Compare string, wet string, and string wet with salt solution. Which of these materials is able to carry electric charges from one can to the other? Which materials don't conduct?

_____

_____

_____

Step 5: Try using other electrically charged objects to make the ferry go. Try a charged sweater, feather, or balloon. Try charged newspaper, table tennis balls, or rubber bands. Some of these things may be held near the can on the wax. Others may be held inside the can or simply dropped into the can. How do they affect the action of the ferry?

_____

_____

_____

Summing Up

Explain the operation of the ferry. What affects how charge is transmitted from can to can?

_____

_____

_____

_____

_____

_____

The Electric Ferry

CONCEPTUAL **Physics**

Activity

Electric Circuits

Let There Be Light

Purpose
To study various arrangements of a battery and bulbs and the effects of those arrangements on bulb brightness.

Required Equipment and Supplies
3 size-D dry-cell batteries
50 cm of bare copper wire
3 flashlight bulbs (1.5 volt)
3 porcelain or plastic bulb sockets

Discussion
Many devices include electronic circuitry, most of which are quite complicated. Complex circuits are made, however, from simple circuits. In this activity you will build one of the simplest yet most useful circuits ever invented — a light bulb!

Procedure
Step 1: Arrange one bulb, one battery, and connecting wire in as many ways as you can to make the bulb emit light. Sketch each of your arrangements, including failures as well as successes. Also describe the similarities between your successful trials.

  1. What do the successful arrangements have in common?

  _____

  _____

Step 2: Using a bulb in a bulb socket (instead of a bare bulb), one battery, and wire, light the bulb in as many ways as you can. Sketch your arrangements and note the ones that work.

  2. With what two parts of the bulb does the holder make contact ?

  _____

  _____

Step 3: Using one battery, light as many bulbs in sockets as you can. Sketch your arrangements and note the ones that work.

  3. If possible, compare your results with those of other students. What arrangements light the most bulbs with only one battery?

  _____

  _____

Let There Be Light

Step 4: Diagrams for electric circuits use symbols such as the ones in Figure A.

Fig. A

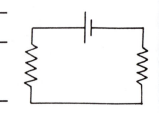

WIRE

BATTERY

LIGHT BULB OR ANY DEVICE THAT USES
ELECTRICAL ENERGY IN A CIRCUIT

Connect the bulbs, battery, and wire as shown in Figure B. Circuits like these are examples of *series* circuits.

4. How does the brightness of the bulbs compare in these two circuits?

Fig. B

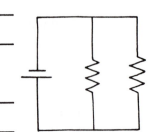

_____

_____

5. What happens if you unscrew a bulb in either circuit?

_____

_____

Step 5: Set up the circuit shown in Figure C. A circuit like this is called a *parallel* circuit.

6. How does the brightness of the bulbs in this circuit compare to that in the series circuit?

Fig. C

_____

_____

7. What happens when either bulb is unscrewed?

_____

_____

Summing Up

8. How do you think most of the circuits in your home are wired— in series or in parallel?

_____

_____

Let There Be Light

CONCEPTUAL **Physics**

Household Circuits

3-Way Switch

Purpose

To explore ways to turn a light bulb on or off from one of two switches.

Required Equipment and Supplies

2.5 volts DC light bulb with sockets
hook-up wire
two 1.5 volt size D batteries with holder, connected in series
two single pole double-throw switches

Discussion

Frequently, multi-story homes have hallways with ceiling lights. It is convenient if you can turn a hallway light on or off from a switch located at either the top or bottom of the staircase. Each switch should be able to turn the light on or off, regardless of the previous setting of either switch. In this acitivity you will see how simple but tricky such a common circuit really is!

Procedure

Step 1: Connect a wire from the positive terminal of a 3 volt battery (two 1.5 volt batteries connected positive to negative) to the center terminal of a single-pole double-throw switch.

Step 2: Connect a wire from the negative terminal of the same battery to one of the light bulb terminals. Connect the other terminal to the center terminal of the other switch.

SINGLE POLE
DOUBLE-THROW SWITCH

Step 3: Now interconnect the free terminals of the switches so that the bulb turns on or off from either switch — regardless of the setting of the other switch.

Step 4: When you succeed, draw a simple circuit diagram of your successful circuit.

Summing Up

1. Will your successful circuit work if you reverse the polarity of the battery?

_____

_____

2. Would your successful circuit operate if the battery were located on the other side of the light bulb?

_____

_____

CONCEPTUAL Physics ————————————————————— | **Activity** |

Magnetic Force

You're Repulsive!

Purpose
To observe the force on an electric charge and the current induced in a conductor moving in a magnetic field.

Required Equipment and Supplies
cathode-ray oscilloscope or Apple II Series computer with monitor
horseshoe magnet
bar magnet
compass
50 cm insulated wire
galvanometer or sensitive milliammeter
masking tape

Discussion
In this activity you will explore the relationship between the magnetic field of a horseshoe magnet and the force that acts on a beam of electrons that move through the field. You will see that you can deflect the beam with different orientations of the magnet. If you had more control over the strength and orientation of the magnetic field, you could use it to "paint" a picture on the inside of a cathode ray tube with the electron beam. This is what happens in a television set.

A changing magnetic field can do something besides make a television picture. It can induce the electricity at the generating station that powers the television set. You will also explore this idea.

Purpose
Step 1: If you are using an oscilloscope, adjust it so that only a spot occurs in the middle of the screen. This will occur when there is no horizontal sweep.

If you are using the monitor of an Apple II Series computer for the cathode ray tube, enter and run the following program to create a spot at the center of the screen:

```
10 HGR: HCOLOR
20 HPLOT 140, 96
```

Caution: Do not leave the dot on the screen longer than needed as it will erode the phosphors at that location. Vary the numbers in the HPLOT X,Y to avoid damaging the monitor.

The dot on the screen is caused by an electron beam that hits the screen.

1. In what direction are the electrons moving?

_____

Step 2: If the north and south poles of your magnet are not marked, use a compass to determine which end is the north pole and label it with a piece of masking tape. (Note: The north pole of a magnet *seeks* the north pole of the earth.)

Place the poles of the horseshoe magnet about 1 cm from the screen. Try the orientations of the magnet shown on the Figure A. Sketch arrows on Figure A to indicate the direction in which the spot moves from the undisturbed location in each case. Try other orientations of the magnet and make sketches to show the direction that the spot moves.

Fig. A

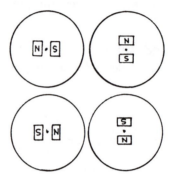

2. Recall that the magnetic field lines outside a magnet run from the north pole to the south pole. The spot moves in the direction of the magnetic force on the beam. How is the direction of the magnetic force on the beam related to the direction of the magnetic field?

_____

_____

Step 3: Aim one pole of a bar magnet directly toward the spot. Record your observations.

_____

_____

Step 4: Change your aim so that the pole of the bar magnet points to the left, right, top, bottom, back and front of the screen. Record your observations.

_____

_____

3. In general, what are the relative directions of the electron beam, the magnetic field, and the magnetic force on the beam for maximum deflection of the electron beam?

_____

_____

_____

Step 5: With a long insulated wire, make a three loop coil that is approximately 5 cm in diameter. Tape the loops together. Connect the ends of the wire to the two terminals of a galvanometer or sensitive milliammeter. Explore the effects of moving a bar magnet in the coil to induce electric current that will cause the galvanometer to deflect. Vary the directions, poles and speeds of the magnet. Also vary the number of loops in the coil, and try magnets of different strengths. Try moving the magnet while keeping the coil stationary and then moving the coil and keeping the magnet stationary. What do you discover?

_____

_____

_____

_____

_____

_____

Summing Up

4. Under what conditions are you able to induce the largest current (the largest deflection of the galvanometer)?

_____

_____

5. Under what conditions can you cause the galvanometer to deflect in the opposite direction?

_____

_____

6. How do the effects of a moving coil in a stationary magnetic field compare to that of a moving magnet through a stationary coil?

_____

_____

You're Repulsive!

CONCEPTUAL Physics━━━━━━━━━━━━━━━━━━━━━━━━━━━━━━━━━ | **Activity** |
Electromagnetic Induction

Jump Rope Generator

Purpose
To demonstrate the operation of a simple generator using a conductor cutting the Earth's magnetic field.

Required Equipment and Supplies
50 foot extension cord (with ground)
galvanometer
two lead wires with alligator clips on at least one end of both wires

Discussion
When a magnetic field is changed in a conducting loop of wire, a voltage and hence a current is induced in the loop. This is what happens when the armature in a generator is rotated. This is what happens when an iron car drives over a loop of wire embedded in the roadway to activate the traffic lights. This is what happens when a piece of wire is twirled like a jumping rope in the earth's magnetic field. Try it and see!

Procedure
Step 1: Attach an alligator clip lead to the ground prong of an extension cord. Attach the other end of the alligator lead to a galvanometer. Jam the other alligator clip (or suitable conductor) into the ground receptacle on the other end of the extension cord. Attach the other end of this lead to the other contact on the galvanometer. Actually, any strand of wire will do — an extension cord is handy and swings well.

Step 2: Align the extension cord in the east/west direction. Leaving about one half of the extension cord on the ground, pick up the middle half and twirl like a jumping rope with another person.

  1. What effect does the rotational speed of the cord have on the deflection of the galvanometer?

_____

Step 3: Repeat with the extension cord aligned in the north/south direction. Observe the difference of the deflection of the galvanometer.

  2. Is it harder to spin the cord in one direction than the other?

_____

  3. What conditions yield the maximum current fluctuations?

_____

Summing Up

4. What does this have to do with Faraday's law?

_____

_____

_____

_____

CONCEPTUAL Physics ——————————————————————————— | **Activity** |

Electric Motors

Workaholic

Purpose
To make a simple dc electric motor.

Required Equipment and Supplies
strong U magnet
power supply
#7 cork, center drilled
12 cm length of 6-mm diameter Pyrex® tubing with rounded ends
push pins or thumbtacks
1 baseboard; about 15 cm x 10 cm x 2 cm
about 3 meters of insulated #30 copper magnet wire

Discussion
It is difficult to imagine a time before electricity. We take it and its marvelous workhorse, the electric motor, for granted. In this activity, you will get an opportunity to not only make a simple motor, but to see how it works as well.

Procedure
Step 1: Begin by carefully inserting a piece of glass tubing (armature shaft) about 10 cm in length through a cork. Make an armature by winding about 20 turns of wire around the cork. Keep the plane of turns parallel to the glass tubing. Leave about 5 cm of wire on each end. To create leads, use a piece of fine sandpaper to remove the enamel insulation from the ends of the wire. Tape the bare leads to the armature shaft with some masking tape on opposite sides of the shaft as shown in Figure A.

Fig. A

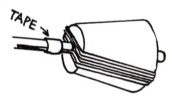
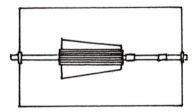

Step 2: Bend two large paper clips so that one end of each clip acts as an axle (Figure B). Mount the armature and shaft with the paper clips and fasten them with push pins or thumbtacks on the baseboard.

Fig. B

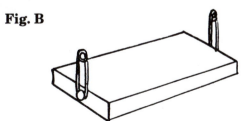

Step 3: Bend a stiff piece of 10 cm bare copper wire "brush" so that it bows towards the armature shaft. Attach the brushes to the board with a push pin or thumbtack (Figure C). The leads from the armature should contact the brushes.

Fig. C

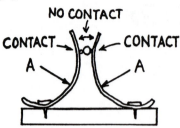
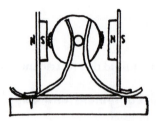

Step 4: Have your partner hold a strong U-magnet directly over the armature without touching it so that the axes of the magnet are perpendicular to the armature shaft. Connect the brush leads to a power supply or dry cell and apply power.

Summing Up

Did your motor work? What happens when you move the magnet away from the armature or vary the voltage?

_____

_____

_____

CONCEPTUAL **Physics**

Experiment

Electirc Current

Ohm Sweet Ohm

Purpose
To investigate how the current in a circuit varies with voltage and resistance.

Required Equipment and Supplies
nichrome wire apparatus with bulb
two-1.5 volt batteries or
two hand-held generators (Genecons preferred)

Discussion
Normally it is desirable that wires in an electrical circuit to stay cool.
There are notable exceptions, however. Nichrome wire is a high resistance
wire capable of glowing red-hot without melting. It is commonly used as
the heating element in toasters, ovens, stoves, hair dryers, etc. In this
experiment, it's used as a variable resistor. Doubling the length of a piece
of wire doubles the resistance; tripling the length triples the resistance, and
so forth.

Tungsten wire is capable of glowing white hot without melting and is used
as the filaments of light bulbs. Light and heat are generated as the current
heats the high resistance tungsten filament. The hotter the filament, the
brighter the bulb. That means a bright bulb has a *lower* resistance than a
dimmer bulb. Just as water flows with more difficulty through a thinner
pipe, electrical resistance is greater for a thinner wire. Manufacturers
make bulbs of different wattages by varying the thickness of the filaments.
So we find that a 100-W bulb has a lower resistance than a 25-W bulb.

The brightness of the bulb can be used as a *current* indicator. Glowing
brightly indicates a large current is flowing through it; dimly lit means a
small current is flowing.

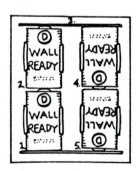

Procedure
Step 1: Connect four D-cell batteries in series, so that the positive
terminal is connected to one negative terminal in a battery holder as shown
in Figure A. This arrangement: with terminal #1 as ground, will provide
you with a variable voltage supply as indicated in Table A.

Fig. A

TERMINAL #'s	VOLTAGE
1-2	1.5
1-3	3.0
1-4	4.5
1-5	6.0

Ohm Sweet Ohm

Fig. B

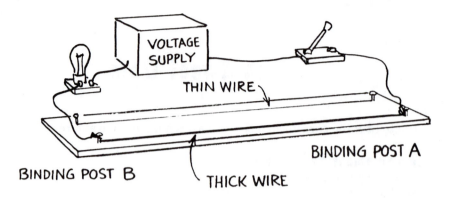

Step 2: Assemble the circuit as shown in Figure B. Label one binding post of the nichrome wire "A" and the other "B". Attach the ground lead (#1) of the voltage supply to one side of a knife switch. Connect the other side of the switch to binding post "A" of the thickest nichrome wire. Connect the 3-volt lead (#3) from the voltage supply to a clip lead of a test bulb. Attach the other clip lead of the test bulb to the other binding post (B) on the nichrome wire.

The voltage supply is now connected so that the current passes through two resistances; the bulb and the nichrome wire. You will vary the resistance in the circuit by moving the clip lead of the test bulb from binding post B to A. Using the standard symbols for the circuit elements, draw a diagram that represents this *series* circuit.

Note: Always apply power from battery packs by closing a switch and make your measurements quickly. Leave the power on just long enough to make your measurements and then open the switch. *Leaving the power on in the circuit for long periods of time will drain your batteries and heat up the wire and filaments in your bulbs and change their resistances.*

Step 3: After carefully checking all your connections, apply power to the circuit by closing the switch. Observe the intensity of the bulb as you move the test bulb lead from binding post B toward A.

1. What happens to the brightness of the bulb as you move it from B to A?

_____

Step 4: Repeat using the thinner nichrome wire. Observe the relative brightness of the bulb as you move the bulb's lead closer to binding post A.

2. How does the brightness of the bulb with the thinner wire compare to the brightness of the bulb when connected to a thicker wire?

_____

3. What effects do the thickness and length of the wire have on its resistance?

_____

_____

_____

Ohm Sweet Ohm

4. Does the current of the circuit increase or decrease as you move the lead closer to binding post B? As you move the lead from B to A, does the resistance of the circuit increase or decrease?

_____

_____

_____

Step 5: Repeat Steps 1 and 2 using the 4.5-volt and 6-volt leads instead of the 3-volt leads.

5. How does the brightness of the test bulb compare for the two nichrome wires using the 4.5 volts instead of 3 volts?

_____

_____

_____

6. Combine your results from Questions 4 and 5. How does the current in the circuit depend upon voltage and resistance?

_____

_____

Going Further

Step 6: Now insert an ammeter into the circuit as illustrated in Figure C. Place the ammeter in series with the voltage supply between terminal (1) of the voltage supply and the switch with the thicker piece of nichrome wire in the circuit. The ammeter will read total current in the circuit. Measure the current in the circuit as you move the test bulb lead from B to A. Be sure to apply power *only* while making the measurements to prevent draining the batteries. Repeat for the thinner wire.

Note: If you are not using a digital meter, you may have to reverse the polarity of the leads if the needle of the meter goes the wrong way (–) when power is applied.

Fig. C

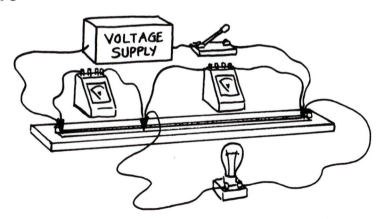

Summing Up

7. Do your results show a decrease in current as the resistance (or length of the wire) is increased?

_____

_____

8. Do your results show an increase in current as the voltage is increased?

_____

_____

9. How does the current in the thicker wire compare when the same voltage is applied to the same length of the thinner wire?

_____

_____

CONCEPTUAL **Physics**

Experiment

Ohm's Law

Voltage Divider

Purpose
To investigate how the resistance of a bulb varies with temperature.

Required Equipment and Supplies
nichrome wire apparatus with bulb
two 1.5-volt batteries

Discussion
As electrical current flows through a wire, the atoms and other electrons provide resistance, just as you would encounter if you tried to run through a crowd of people. Some of the electrons' energy is dissipated as heat as the electrons jostle one another in the wire. The greater the interaction between the current and the atomic lattice that makes up the wire, the greater the energy dissipated as heat. Tungsten is used as the filament wire in light bulbs because it has these physical properties: high resistance, high melting point, and relatively low chemical reactivity. In this experiment, you will be able to measure how the resistance of the bulb varies with temperature.

Procedure
Step 1: Place a ruler underneath the nichrome wire and tack the ruler down with masking tape. Assemble the circuit as shown in Figure A. Label one binding post of the nichrome wire "A" and the other "B". Attach the ground lead (#1) of the voltage supply to one side of a knife switch. Connect the other side of the switch to binding post A of the thicker nichrome wire. Connect the 3-volt lead (#3) from the voltage supply to binding post B. Attach one clip lead of the test bulb to binding post (B) and the other to binding post A. The voltage supply is now connected so that the current splits into two branches; one through the bulb and the other through the nichrome wire. Using the standard symbols for the circuit elements, draw a diagram that represents this *series* circuit.

Fig. A

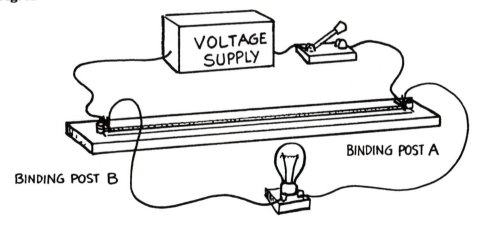

Fig. B

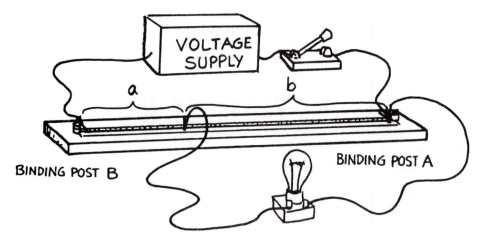

As you move the clip lead of the test bulb from binding post B to A you vary the amount of current that flows through each branch. This circuit is frequently encountered in electronics. Let's take a look at what's happening in more detail. The total current passes from the voltage supply through the wire segment "a" in Figure B. As you move the lead from the bulb towards the other binding post, the total current in the circuit splits into two branches. One branch flows through the bulb; the other flows through segment "b" of the nichrome wire. The branch currents combine at the end of the nichrome wire and return to the voltage supply to complete the circuit. Because the voltage drops as you move the lead towards binding post B, less voltage is applied across the bulb. Since the circuit "divides" the voltage between the remaining wire segment and the bulb, it is commonly known as a "voltage divider".

Use the appropriate symbols and draw a diagram that represents this *compound* circuit. A compound circuit is a combination of series and parallel circuits.

Fig. C

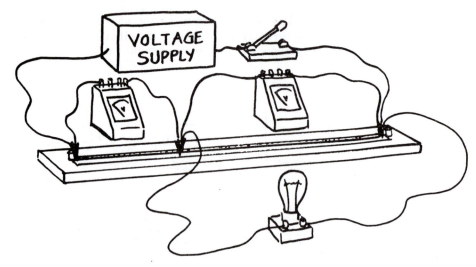

Voltage Divider

Step 2: Measure the voltage across segment "a" of the nichrome wire and the bulb starting with the bulb leads connected to each binding post ($L = 0$). Do this by number connecting the voltmeter in parallel as shown in Figure C. Then move the bulb lead from binding post B to a point $\frac{1}{3}$ the length ($l = \frac{1L}{3}$) of the wire from the binding post B. Measure the voltage across segment "a" and across the bulb. Note that the voltage across segment "b" is the same as the voltage across the bulb. Record your measurements in Table A. Now move the bulb lead to a point two-thirds the length of the wire ($l = \frac{2L}{3}$) and measure the voltages.

Table A

DISTANCE	VOLTAGE (V)	CURRENT (A)	RESISTANCE (Ω)
0 L			
⅓ L			
⅔ L			
L			

1. What happens to the voltage as you move the bulb lead closer to binding post B? How are the voltage drops across segment "a" and the bulb related?

_____

_____

_____

Voltage Divider

Fig. D

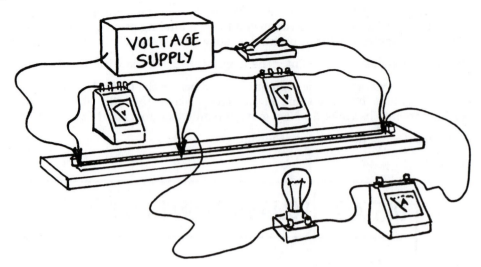

Step 3: Repeat, except this time measure the current through the bulb as well as the voltage across it. Do this by connecting an ammeter in series with the bulb as shown in Figure D. This will enable you to compute the resistance of the bulb ($\frac{V}{I}$) as it dims. Record your results in Table B.

Table B

DISTANCE	VOLTAGE (V)	CURRENT (A)	RESISTANCE (Ω)
0 L			
⅓ L			
⅔ L			
L			

Voltage Divider

Summing Up

2. Compute the resistance, $\frac{V}{I}$ of the bulb, in the four different positions. Does the resistance of the bulb increase or decrease with brightness?

_____

_____

_____

3. Is the change in resistance significant? Can you explain why?

_____

_____

_____

4. Do you think engineers must take this into account when designing light bulbs?

_____

_____

_____

Voltage Divider

Name: _____ Section: _____ Date: _____

Cranking-Up Qualitatively

Purpose
To observe the work done in a series circuit compared to a parallel circuit.

Required Equipment and Supplies
hand crank generator (Genecon)
parallel bulb apparatus

Discussion
In this activity you will not only see some differences between series and parallel circuits but feel them as well.

Procedure
Step 1: Assemble four bulbs in a series configuration as shown in Figure A. Screw all the bulbs into their sockets. Connect the sockets with clip leads or wires.

Fig. A   **Fig. B**

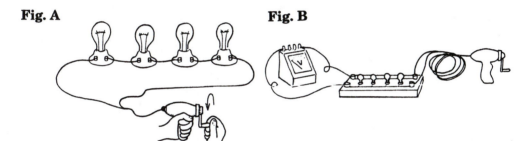

Step 2: Connect one lead of a Genecon hand-cranked generator to one end of the string of bulbs and the other lead to the other end of the string. Crank the Genecon so that all the bulbs light up. Now disconnect one of the bulbs from the string and re-connect the Genecon. Crank the Genecon so that the three remaining bulbs are energized to the same brightness as the four-bulb arrangement. How does the crank *feel* now? Repeat removing one bulb at a time, comparing the cranking torque each time.

Step 3: Assemble the circuit with the parallel-bulb apparatus as shown in Figure B. Each end of the bulb apparatus has two terminals. Connect the lead of a voltmeter to one pair of terminals on one end of the apparatus.

Step 4: Connect the leads of the Genecon to the terminals on the other end of the apparatus. Crank the Genecon with all the bulbs unscrewed in the sockets so they don't light. Then have your partner screw them in one at a time as you crank on the Genecon. Try to keep the bulbs energized at the same brightness as each bulb is screwed into its socket.

Cranking-Up Qualitatively

Summing Up

1. What do you notice about the *torque* required to crank the Genecon at a constant speed as more bulbs were added to the circuit?

_____

_____

2. How would you compare the amount of torque required to crank the Genecon to energize four bulbs in series with the torque to energize to four bulbs in parallel?

_____

_____

3. If all the bulbs in series and parallel circuits are glowing equally brightly, is the energy expended (required torque on the crank of the Genecon) the same?

_____

_____

Cranking-Up Qualitatively

CONCEPTUAL **Physics** ▬▬▬▬▬▬▬▬▬▬▬▬▬▬▬ ██ Experiment ██

Electric Motors and Generators

Cranking-Up Quantitatively

Purpose
To investigate the power consumed in a series circuit compared to that in a parallel circuit.

Required Equipment and Supplies
hand crank generator (Genecon preferred)
parallel bulb apparatus
voltmeter
ammeter

Discussion
Now, we are going to repeat "Cranking-Up Qualitatively" in a quantitative fashion using a voltmeter and ammeter.

Procedure
Step 1: Assemble four bulbs in a series circuit and connect the meters as shown in Figure A. Connect the voltmeter in parallel with the bulbs so you can measure the total voltage applied to the circuit as well as the voltage across each bulb. Connect the 3-volt lead from the voltage supply to one terminal of the bulbs and the ground connection to one lead of an ammeter. Connect the other lead of the ammeter to the second terminal of the bulbs. The ammeter will measure the *total* current in the circuit.

Fig. A

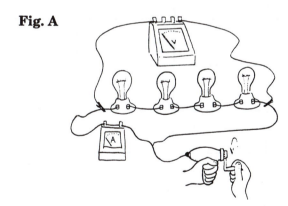

Note: If you are *not* using digital meters, you may have to reverse the polarity of the leads if the needle of the meter goes the wrong way (–) when power is applied.

Close the switch and apply power to the circuit and then measure:

  a) The current in the circuit.
  b) The voltage applied to the circuit.
  c) The voltage across each bulb.

Now remove one of the bulbs from the string and repeat your measurements and repeat for three, two, and then for one bulb. Record your data in Table A.

Table A

# OF BULBS	TOTAL CURRENT (A)	TOTAL VOLTAGE (V)	VOLTAGE ACROSS EACH BULB (V)			

Step 2: Repeat using the 4.5-volt terminal of the voltage supply instead of the 3-volt terminal. Record your data in Table B.

Table B

# OF BULBS	TOTAL CURRENT (A)	TOTAL VOLTAGE (V)	VOLTAGE ACROSS EACH BULB (V)			

Summing Up

1. Do you observe any change in brightness as the number of bulbs in the circuit changes?

_____

2. Does the voltage applied in the circuit change as you add more bulbs?

_____

Cranking-Up Quantitatively

3. How are the voltages across each bulb related to the voltage applied in the circuit?

_____

4. How does the current in the battery change when more bulbs are added?

_____

5. Did any of the relationships you discovered between voltages and currents change when you applied 4.5 volts instead of 3 volts?

_____

Step 3: Assemble the circuit and connect the meters as shown in Figure B. Connect the voltmeter in parallel with the bulbs by connecting the voltmeter to two terminals on one end of the parallel-bulb apparatus. Connect the 3-volt lead from the voltage supply to one terminal of the parallel-bulb apparatus. Connect the ground lead from the voltage supply to one lead of an ammeter; connect the other lead of the ammeter to the second terminal of the parallel-bulb apparatus. The ammeter will measure the *total* current in the circuit.

Fig. B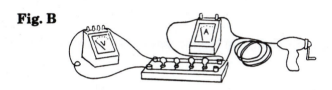

Make sure the bulbs are not loose in their sockets. Close the switch and apply power to the circuit. Observe their brightness. Now unscrew the bulbs one at a time.

Step 4: Screw the bulbs back in, one at a time, each time measuring:

a) The current in the circuit.
b) The voltage applied to the circuit.
c) The voltage across each bulb.

Record your data in Table C.

Table C

# OF BULBS	TOTAL CURRENT (A)	TOTAL VOLTAGE (V)	VOLTAGE ACROSS EACH BULB (V)			

Cranking-Up Quantitatively

Step 5: Repeat Steps 3 and 4 using the 4.5-volt terminal of the voltage supply instead of the 3-volt terminal. Record your data in Table D.

Table D

# OF BULBS	TOTAL CURRENT (A)	TOTAL VOLTAGE (V)	VOLTAGE ACROSS EACH BULB (V)			

Summing Up

6. Do you observe any change in brightness as the number of bulbs in the circuit changes?

_____

7. Does the voltage across each bulb change as more bulbs are added to or subtracted from the circuit?

_____

8. Does the applied voltage to the circuit change as you add more bulbs?

_____

9. How does the current in the battery change as the number of bulbs in the circuit changes?

_____

10. Did any of the relationships you discovered between voltage and current change when you applied 4.5 volts instead of 3 volts?

_____

CONCEPTUAL Physics ━━━━━━━━━━━━━━━━━━━━━━━━━━━━━
Electric Motors and Generators

Motors and Generators

Purpose
To observe the effects of electromagnetic induction.

Required Equipment and Supplies
2-hand-cranked generators (Genecons preferred)
3-6 volt power supply
voltmeter (digital or analog)
stopwatch
2 large demonstration U-magnets
2 coils of magnet wire

Discussion
Generators and motors operate are similar devices with input and output reversed. Both involve wire loops in a magnetic field. The input in a motor is electric current, which is deflected as it enters the magnetic field inside the device. This deflection turns the wire loop and mechanical energy is the output. In a generator, the wire loop is forced to rotate by mechanical means. Initially static charges in the wire are deflected to produce an electirc-current output.

In a generator, the wire loop is rotated through a magnetic field. The wire loop provides a conducting path for the charges as they are deflected at right angles by the magnetic field.

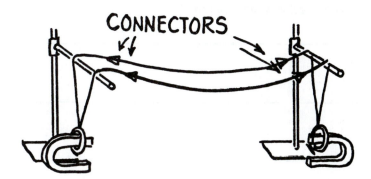

Procedure
Step 1: Set up the magnets and the wires as shown. Move one coil to one side and then back again.

  1. What happens to the other coil? Explain.

_____

_____

2. Does the same thing happen if the coils are not connected to each other?

_____

_____

3. What happens if the poles of the other magnet are reversed?

_____

_____

Step 2: Study the construction of a Genecon. Note that it consists of a dc motor whose armature shaft is connected to the hand crank via gears. Adjust the range of a dc power supply so that voltage output is a maximum of 5 volts. Attach the leads of the Genecon to the power supply. Hold the Genecon so that its handle is free to rotate. Slowly increase the voltage of the power supply and observe the operation of the Genecon. Reverse the polarity of the leads and repeat. What do you observe?

_____

_____

Step 3: Connect the leads of one Genecon to the leads of another. Have your partner hold one Genecon while you turn the crank on the other.

4. What happens to the other Genecon when you turn the crank on yours:

a) in the clockwise direction ?

_____

b) in the counterclockwise direction?

_____

c) fast ?

_____

d) slowly?

_____

e) with the leads reversed?

_____

Summing Up
5. Does the Genecon behave any differently when connected to another Genecon rather than a power supply? What is the principal difference is this case ?

_____

_____

Going Further—Efficiency

Step 4: Attach the leads of the Genecon to a voltmeter whose range is set to a maximum of 10 volts. Crank the handle at various speeds. Observe the readings on the voltmeter.

Attach the leads of a Genecon to the terminals of a single bulb in a socket. Attach the leads of a voltmeter to the same terminals of the bulb. The voltmeter should be adjusted so that the full-scale reading is 5 or 10 volts. Turn the handle of the Genecon at a steady fixed speed so that the voltmeter reads 0.5 volts. Have your partner time you for ten seconds as you count the number of turns of the crank. Record your data in Table A.

Table A

NUMBER OF TURNS	VOLTAGE

Repeat this procedure up to 4 volts in $\frac{1}{2}$-volt increments. Observe the intensity of the light bulb.

Step 5: The speed the armature moves through the magnetic field inside the generator is proportional to the number of turns per second. Make a graph of Voltage (volts) vs. Speed (turns/s) using your data.

Summing Up

6. How does the intensity of the lamp vary as you increase the crank speed?

_____

7. How does the voltage of the generator vary with the speed of the armature through the magnetic field?

_____

8. What is the optimum speed (that is, the speed after which there is very little increase in voltage with increasing crank speed) to generate voltage using this generator?

_____

9. Where does the energy to light the bulb originate?

_____

Going Even Further—Increasing Your Capacity

A capacitor is an electronic device that stores electric charge and electric energy. The flash attachment for cameras uses a capacitor to store the energy needed to provide a sudden flash of light. Capacitors are used to smooth out ac ripples in a direct current voltage supply such as those used to power your calculator or radio.

Step 6: Attach the leads of a Genecon to a large capacitor—a 1.0 farad capacitor, if available. Crank the Genecon briskly for a half minute or so. Let go of the crank.

10. Which way does it rotate — in the same direction you had been cranking or in the opposite direction? Are you surprised? Why? Can you explain why it rotated the way it did?

_____

_____

_____

_____

_____

Motors and Generators

CONCEPTUAL **Physics**

Pinhole Lens

Lensless Lens

Purpose
To investigate the operation of a pinhole lens and compare it to the eye.

Required Equipment and Supplies
3" x 5" card
straight pin
meterstick

Discussion
Both a prism and a lens deviate light because their faces are not parallel (only at the center of a lens are both faces parallel to each other). As a result, light passing through the center of a lens undergoes the least deviation. If a pinhole is placed at the center of the pupil of your eye, the undeviated light forms an image in focus no matter where the object is located. Pinhole vision is remarkably clear. In this activity, you will use a pinhole to enable you to see fine detail more clearly.

Procedure
Step 1: Bring this printed page closer and closer to your eye until you cannot clearly focus on it any longer. Even though your pupil is small, your eye does not act like a true pinhole camera because it does not focus well on nearby objects.

Step 2: Poke a single pinhole about one centimeter from the edge of a piece of card (like a 3" x 5" card). Hold the card in front of your eye and read these instructions through the pinhole. Bright light on the print may be required. Bring the page closer and closer to your eye until it is a few centimeters away. You should be able to read the type clearly. Then quickly remove the card and see if you can still read the instructions without the benefit of the pinhole.

Summing Up
1. Did the print appear magnified when observed through the pinhole?

_____

_____

2. Did the pinhole actually magnify the print?

_____

_____

3. Why was the page of instructions dimmer when seen through the pinhole than when seen using your eye alone?

_____

_____

4. A nearsighted person cannot see distant objects clearly without corrective lenses. Yet such a person can see distant objects clearly through a pinhole. Explain how this is possible. (And if you are nearsighted yourself, try it and see!)

_____

_____

CONCEPTUAL **Physics**

Pinhole Camera

Camera Obscura

Purpose
To observe images formed by a simple convex lens and compare cameras with and without a lens.

Required Equipment and Supplies
covered shoebox approximately 4" x 6" x 12" with a pinhole and a 25 mm converging lens set in one end with glassine paper in the middle of the box, (as shown in Figure A)
aluminum foil
masking tape

Discussion
The first camera, known as a *camera obscura*, used a pinhole opening to let light in. The light that passes through the pinhole forms an image on the inner back wall of the camera. Because the opening was small, a long time was required to expose the film sufficiently. A lens allows more light to pass through and still focus the light onto the film. These cameras require much less time for exposure, and the pictures came to be called "snapshots".

Procedure
Step 1: Use the camera constructed as in Figure A. Tape some foil over the lens of the box so that only the pinhole is exposed. Hold the camera towards a brightly illuminated scene, such as the window during the daytime. Light enters the pinhole and falls on the glassine paper. Observe the image of the scene on the glassine paper.

Fig. A

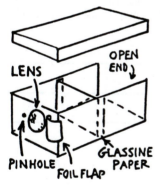

1. Is the image on the screen upside down (inverted)?

_____

2. Is the image on the screen reversed left to right?

_____

Step 2: Now seal off the pinhole with a piece of masking tape and open the foil door to allow light through the lens. Move the camera around. You can watch people or cars moving.

3. Is the image on the screen upside down (inverted)?

_____

4. Is the image on the screen reversed left to right?

_____

Step 3: Unlike a lens camera, pinhole cameras focuses equally well on objects at practically all distances. Aim the camera lens at a nearby objects and see if the lens focuses on nearby objects.

5. Does the lens really focus on nearby objects as well as it does on distant ones?

_____

Step 4: Draw a ray diagram as follows. First, draw a ray for light that passes from the top of a distant object through a pinhole and onto a screen. Second, draw another ray for light that passes from the bottom of the object through the pinhole and onto the screen. Then sketch the image created on the screen by the pinhole.

Summing Up

6. Why is the image created by the pinhole dimmer than the one created by the lens?

_____

_____

_____

7. How is a pinhole camera similar to your eye? Do you think that the images formed on the retina of your eye are upside down? Your explanation might include a diagram.

_____

_____

_____

CONCEPTUAL **Physics**

Virtual Images

Mirror, Mirror, on the Wall . . .

Purpose
To investigate the relationship between the size of your image and your distance from a plane mirror.

Required Equipment and Supplies
large mirror, preferably full length
ruler
4" x 5" mirror
masking tape

Discussion
Why do shoe stores and clothier shops have full-length mirrors? To facilitate seeing your toes, right?

Procedure
Step 1: Stand at arm's length in front of a vertical full-length mirror. Stand straight and stare right ahead and smile! Reach out and place a small piece of masking tape on the image of the top of your head. Now stare at your toes. Place the other piece of tape on the mirror where your toes are seen. Use a meterstick to measure the distance from top of your head to your toes. How does the distance between the pieces of tape on the mirror compare to your height?

_____

Step 2: Now stand about 3 meters from the mirror and repeat. Stare at the top of your head and toes and have an assistant move the tape so that the pieces of tape mark where head and feet are seen. Move further away or closer, and repeat. What do you discover?

_____

Summing Up
1. Does the location of the tape depend on your distance from the mirror?

_____

2. What is the shortest mirror in which you can see your entire image? Do you *believe* it?

_____

_____

_____

Mirror, Mirror, on the Wall

Going Further

Try this one if a full-length mirror is not readily available *or* you are a disbeliever! Hold a ruler next to your eye. Measure the height of a common pocket mirror. Hold the mirror in front of you so that the image includes the ruler. How many centimeters of the ruler appear in the image? How does this compare to the height of the mirror?

_____

_____

_____

_____

_____

CONCEPTUAL **Physics**
Refraction of Lenses

Activity

Air Lens?

Purpose
To apply your knowledge of light behavior and glass lenses to a different type of lens system.

Required Equipment and Supplies
2 depression microscope slides
light source
screen

Discussion
Ordinary lenses are made of glass. A glass lens that is thicker at the center than at the edge is convex in shape, converges light, and is called a *converging* lens. A glass lens that is thinner at the middle than at the edge is concave in shape, diverges light, and is called a *diverging lens*. But what about two concave depressions of glass pressed together to form a convex air pocket?

Procedure
Step 1: An air wedge encased in glass can be produced by placing two depression microscope slides together as shown in Figure A. Does this arrangement behave as a diverging or converging lens? Make a ray diagram to help explain your answer.

Fig. A

AIR LENS

Step 2: Use your lens with a light source and screen to check your prediction. What do you discover?

1. Why is the statement "the shape of a lens determines whether it is a converging or diverging lens" not always true?

_____

_____

2. Draw a ray diagram using a concave air lens encased in glass to show what this lens does to light rays passing through it. Diverging or converging lens?

Air Lens?

Summing Up

3. Why is it misleading to call the air pocket a *lens*?

_____

_____

_____

_____

_____

_____

Air Lens?

CONCEPTUAL **Physics** | **Activity** |
Formation of Images

Images

Purpose
To formulate ideas about how reflected light travels to your eyes.

Required Equipment and Supplies
2 small plane mirrors
supports for the mirrors ("bulldog" spring clips work well)
2 single-hole rubber stoppers
2 pencils
2 sheets of paper

Discussion   **Fig. A**
Reflections are interesting. Reflections of reflections are fascinating.
Reflections of reflections of reflections are . . . you will see for yourself in this
activity!

Purpose
Step 1: Place the pencils in the rubber stoppers and set one plane mirror
upright in the middle of a sheet of paper, as shown in Figure A. Stand one
pencil vertically in front of the mirror. Hold your eye steady at the height of
the mirror. Locate the image of the pencil in the mirror. Place the second
pencil where the image of the first appears to be. If you have located the image
correctly, the image of the first pencil and the second pencil itself will remain
"together" as you move your head from side to side.

    1. How does the distance from the first pencil to the mirror compare to the distance of
the mirror to the image?

_____

_____

Step 2: Sketch the paths you think "rays" of light take going from the first pencil to
your eye as you observe the image. Draw a dotted line to where the image appears to
be located as seen by the observing eye.

Step 3: Set two mirrors upright and at right angles to each other in the middle of a second sheet of paper. Place a pencil in its stopper between the mirrors, as in Figure B.

2. How many images do you see?

_____

Step 4: Make a sketch of where the images are located. Draw the paths you think the light takes as it goes from the pencil to your eye.

Step 5: Decrease the angle between the two mirrors. Observe the images created by the reflections. Try several angles such as 30°, 45°, 60°, etc.

Summing Up

3. What happens to the number of images you get when you decrease the angle between the two mirrors?

_____

_____

_____

Images

CONCEPTUAL Physics   █ Experiment █

Pinole Image Formation

Sunballs

Purpose
To estimate the diameter of the sun.

Required Equipment and Supplies
small piece of cardboard
sharp pencil or pen
meterstick

Discussion
Take notice of the round spots of light on the shady ground beneath trees.
These are sunballs — images of the sun (Figures 1.6, 1.7, and 1.8 in
Conceptual Physics). They are cast by openings between leaves in the trees
that act as pinholes. The diameter of a sunball depends on its distance from
the small opening that produces it. Large sunballs, several centimeters in
diameter or so, are cast by openings that are relatively high above the ground,
while small ones are produced by closer "pinholes." The interesting point is
that the ratio of the diameter of the sunball to its distance from the pinhole is
the same as the ratio of the sun's diameter to its distance from the pinhole.

Since the sun is approximately 150,000,000 km from the pinhole, careful measurement
of this ratio tells us the diameter of the sun. That's what this activity is all about.
Instead of finding sunballs under the canopy of trees, make your own easier-to-measure
sunballs.

Procedure
Poke a small hole in a piece of cardboard with a pen or sharp pencil. Hold the
cardboard in the sunlight and note the circular image that is cast on a
convenient screen of any kind. This is an image of the sun. Note that its size
does not depend on the size of the hole in the cardboard (pinhole), but only on
its distance from the pinhole to the screen. The greater the distance between
the image and the cardboard, the larger the sunball.

Position the cardboard so the image exactly covers a nickel, dime or something
that can be accurately measured. Carefully measure the distance to the small
hole in the cardboard.

   diameter of the sunball
   distance to the pinhole

With this ratio, estimate the diameter of the sun. Show your calculations.

       diameter of the sun, d = _____

Summing Up

1. Will the sunball still be round if the pinhole is square shaped? Triangular shaped? (Experiment and see!)

_____

_____

_____

_____

2. If the sun is low so the sunball is elliptical, should you measure the small or the long width of the ellipse for the sunball diameter in your calculation of the sun's diameter? Why?

_____

_____

_____

3. If the sun is partially eclipsed, what will be the shape of the sunball?

_____

_____

_____

Sunballs

CONCEPTUAL **Physics** | **Experiment**

Real and Virtual Images

Pepper's Ghost

Purpose
To explore the formation of mirror images by a plate of glass.

Required Equipment and Supplies
two candles of equal size, in holders
1 thick plate of glass, approximately 30 cm x 30 cm by 1 cm (the thicker glass makes the double image easier to see; thinner glass is more breakable but OK)
2 supports for the glass plate
matches

Discussion
John Henry Pepper (1821-1900), a chemistry professor in London, used his knowledge of image formation to perform as an illusionist. One of his most impressive illusions is because glass both *reflects* and *transmits* light.

Procedure
Step 1: Light one candle, and place it about 5 cm in front of a vertical glass plate. Place a similar unlighted candle behind the glass at the position where the flame of the lighted candle appears to be on the unlighted candle.

1. How does the distance from the lighted candle to the glass plate compare to the distance from the glass plate to the unlighted candle?

_____

Step 2: Look carefully at an angle to the glass, and you should see a double image of the candle flame.

2. Can you explain the double image?

_____

_____

Step 3: Look at the glass plate from the side, at a small angle with the surface of the glass. You will see three or more "ghost" images of the candle flame.

Pepper's Ghost

Summing Up

3. Explain how these "ghost" images are formed. You may want to include a diagram with your explanation.

_____

_____

_____

_____

Pepper's Ghost

Name: _____ Section: _____ Date: _____

Experiment

Kaleidoscope

Purpose
To apply the principles of reflection to a mirror system with multiple reflections.

Required Equipment and Supplies
two 4"x 5" plane mirrors
masking tape
clay
viewing object
protractor
toy kaleidoscope (optional)

Discussion
Have you ever held a mirror in front of you and another mirror in back of you in order to see the back of your head? Did what you see surprise you?

Procedure
Step 1: Hinge the two mirrors together with masking tape to allow them to open at various angles. Use clay and a protractor to hold the two mirrors at an angle of 72°. Place the object to be observed inside the angled mirrors. Count the number of images resulting from this system and record in Table A.

Table A

Step 2: Reduce the angle of the mirrors by 5° at a time, and count the number of images at each angle. Record your findings in Table A.

As the angle between the mirrors becomes closer to 0°, the number of images increases enormously. Two mirrors placed face to face could have an unlimited number of images as the reflection is bounced back and forth. You'll see this whenever you are between two parallel mirrors, like in a barber shop or beauty salon. Physics is everywhere!

ANGLE	NUMBER OF IMAGES
72°	
67°	
62°	
57°	
52°	
47°	
42°	
37°	
32°	
27°	

Step 3: If one is available, study and observe the operation of a toy kaleidoscope. Explain how it works.

_____

_____

_____

_____

_____

Summing Up

1. Explain the reason for the multiple images you have observed.

_____

_____

_____

_____

2. What effect does the angle between the mirrors have on the number of images?

_____

_____

3. Using the information you have gained, explain the construction and operation of a toy kaleidoscope.

_____

_____

_____

_____

_____

Multiple Reflections

CONCEPTUAL **Physics** ────────────────────────────────── | **Activity** |

Half-Life

Half of a Half

Purpose
To develop an understanding of half-life and radioactive decay.

Required Equipment and Supplies
shoe box
200 pennys (or equivalent)
graph paper

Optional
computer
Data Plotter software or equivalent

Discussion
Many things grow at a steady rate, such as population, money in a savings account, and the accumulated thickness of a sheet of paper that is continually folded over onto itself (see Appendix IV in *Conceptual Physics*). Many other things decrease at a steady rate, like the value of money in the bank, charge on a discharging capacitor, and the amount of certain materials during radioactive disintegration. A useful way to describe the rate of decrease is in terms of *half life*—the time it takes for the quantity to reduce to half its value. For steady decrease, called "exponential" decrease, the half life stays the same.

Radioactive materials are characterized by their rates of decay and are rated in terms of their half lives. You will explore this idea in this activity.

Procedure
Step 1: Place more than 100 pennies in the shoe box and place the lid on the box. Shake the box for several seconds. Open the box and remove all the pennies head-side up. Count these and record the number in Table A. Do not put the removed pennies back in the box.

Table A

	TOTAL PIECES				
SHAKE NUMBER	NUMBER OF PIECES REMOVED	NUMBER OF PIECES REMAINING	SHAKE NUMBER	NUMBER OF PIECES REMOVED	NUMBER OF PIECES REMAINING

Step 2: Repeat Step 1 until one or no coins remain. Record the number removed each time.

Step 3: Add the total coins removed to find the original number of coins. Now subtract the number of coins removed each time from the total to find the coins remaining after each shake.

Step 4: Now graph the Coins Remaining (vertical axis) vs. Number of Shakes (horizontal axis). Plot the data and draw a smooth line that best fits the points.

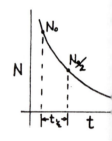

Summing Up

1. What is the meaning of the graph you obtained?

_____

_____

_____

2. Approximately what percent of the remaining coins were removed on each shake? Why?

_____

_____

_____

3. Each shake represents a "half-life" for the coins. What is meant by half-life?

_____

_____

_____

Going Further

Plot your data using *Data Plotter*. Try to discover a way to make your graph come out a straight line. Ask your instructor for assistance if you need help.

Instead of using pennies, try using dice and remove all "3's" (or any other number) etc., and determine the half-life. Before actually doing this, predict how the half-life compares to the activity of the pennies.

Half of a Half

CONCEPTUAL **Physics** | **Activity**

Chain Reaction

Chain Reaction

Purpose
To simulate a simple chain reaction

Required Equipment and Supplies
100 dominoes
large table or floor space
stopwatch

Discussion
Give your cold to two people who in turn each give it to two
others who in turn do the same on down the line and before you
know it everyone in class is sneezing. You have set off a chain
reaction. Similarly, when one electron in a photomultiplier tube
in certain electronic instruments hits a target that releases two
electrons that in turn do the same on down the line, a tiny input
produces a large output. When one neutron triggers the release
of two or more neutrons in a piece of uranium, and the triggered
neutrons trigger others in succession, the results can be
devastating (Chapter 33, *Conceptual Physics*). In this activity
you'll explore this idea.

Fig. A

Procedure
Step 1: Set up a strand of dominoes about half a domino length apart in a straight
line. Gently push the first domino over, and measure how long it takes for the entire
strand to fall over. Try to ascertain whether the rate the dominoes fall along the line
increases, decreases, or remains about the same.

Step 2: Arrange the dominoes as in Figure A, so that when one domino falls, another
one or two will be knocked over. These dominoes then knock over more dominoes so
that the reaction will grow. Set up until you run out of dominoes or table space. When
you finish, push the first domino over and watch the reaction. Notice the number of the
falling dominoes per second at the beginning versus the end.

Summing Up
1. Which reaction took a shorter time to knock over all the dominoes?

_____

_____

2. How did the number of dominoes being knocked over per second change in each
reaction?

_____

_____

3. What made each reaction stop?

_____

_____

4. Now imagine that the dominoes are the neutrons released by uranium atoms when they fission (split apart). Neutrons from the nucleus of a fissioning uranium atom hit other uranium atoms and cause them to fission. This reaction continues to grow if there are no controls. Such an uncontrolled reaction occurs in a split second and is called an *atomic explosion*. How is the domino chain reaction in Step 2 similar to the atomic fission process?

_____

_____

_____

5. How is the domino reaction in Step 2 dissimilar to the nuclear fission process?

_____

_____

_____

Chain Reaction

CONCEPTUAL **Physics** | Experiment

Nuclear Model of the Atom

Nuclear Marbles

Purpose

To determine the diameter of a marble by indirect measurement.

Required Equipment and Supplies

several metersticks
10 marbles

Optional
Apple II computer
"Nuclear Marbles" software or equivalent

Discussion

People sometimes have to resort to something besides a sense of sight to determine the shape and size of things, especially for things smaller than the wavelength of light. One way to do this is to fire particles at the object to be investigated, and study the paths of deflection of the particles that bounce off the object. Ernest Rutherford inferred the size of the nuclei of gold atoms by studying how alpha particles were deflected by the nuclei in gold foil (Chapter 31, *Conceptual Physics*). In this activity, you will determine the diameter of a marble by rolling other marbles at it.

You are not allowed to hold a meterstick across the target marble to measure it directly, at least not in the first part of the activity. You will roll other marbles or spheres at the target marble or "nucleus" and determine its size from the ratio of collisions to trials. It's a little bit like throwing snowballs while blindfolded at a car. If you have very few hits per certain number of trials then the car will "feel" small.

First, use a bit of reasoning to arrive at a formula for the diameter of the nuclear marbles (NM). Then at the end of the experiment you can measure the marbles directly and compare your results.

When you roll a marble toward a nuclear marble, there is a certain probability of a hit between the rolling marble (RM) and the nuclear marble (NM). One expression of the probability P of a hit is the ratio of the path width required for a hit to the width L of the target area (see Figure A). The path width is equal to two RM radii plus the diameter of the NM, as shown in Figure B. The probability P that a rolling marble will hit a lone nuclear marble in the target area is

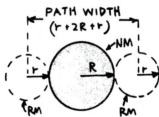

$$P = \frac{\text{path width}}{\text{target width}}$$

$$= \frac{(2R + 2r)}{L} = \frac{2(R + r)}{L}$$

where R = the unknown radius of the NM
  r = the radius of RM
  $R+r$ = the distance between the centers of RM and NM that are touching
and L = the width of the target area.

If the number of target marbles is increased to N, the probability of a hit is increased by a factor of N. Thus, the probability that a rolling marble will hit one of the N nuclear marbles is

$$P = \frac{2N(R+r)}{L}$$

The probability of a hit can also be determined experimentally by the ratio of the number of hits to the number of trials.

$$P = \frac{H}{T}$$

where H = the number of hits
and T = the number of trials.

You now have two expressions for the probability of a hit. Assume that both of the expressions are equivalent. If the radius of the rolling marble and nuclear marble are equal, then $R+r = d$, where d is the diameter of either one of the marbles. By using the two equations for P, try to write an expression for d in terms of $H, T, N,$ and L.

marble diameter, $d =$ _____

Fig. A

This is the formula you are going to test.

Procedure

Step 1: Place 6 to 9 marbles in an area 60 cm wide ($L = 60$ cm) as in Figure A. Roll additional marbles randomly toward the whole target area from the release point. If a RM hits two nuclear marbles, it counts as just one hit. If an RM goes outside the 60 cm area, don't count that trial. A significant number of trials need to be made — at least 200 — before the results become statistically significant. Record your total number of hits H and total number of trials T.

$H =$ _____

$T =$ _____

Step 2: Use your formula from the *Discussion* to find the diameter of the marble. Show your computations.

estimated marble diameter, $d =$ _____

Nuclear Marbles

Step 3: Measure the diameter of one marble.

measured diameter, d = _____

Going Further
Step 4: If a computer is available, you can do this activity using the "Nuclear Marbles" program. Enter your data and compare your value with that of the computer's.

estimated marble diameter, d = _____

computer's value for diameter, d = _____

Summing Up
1. Compare your result for the computed diameter using the formula with your direct measurement of the marble's diameter. What is the percentage difference between the computed and measured values?

percentage difference = _____

2. State a conclusion you can draw from this experiment.

_____

_____

_____

_____

_____

Nuclear Marbles

CONCEPTUAL **Physics**

Vector Addition

Activity

Vector Walk

Purpose
To see that resultant displacement does not depend on the order components are added.

Required Equipment and Supplies
12 popsicle sticks; 3 or 4 short, the rest long
brown bag
masking tape

Discussion
You've just arrived in San Francisco for a physics teachers', conference. You're staying at a hotel downtown, and you would like go the *Green House* for Sunday brunch. The hotel clerk gives you directions after you explain that you would like to go for nice long walk and end up at the *Green House*. On the way out you think it wise to double check yourself, so you ask the taxi cab driver for directions. They are completely different. Now what do you do? Are the directions the clerk gave correct? What about the cab driver's?

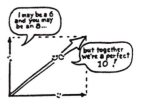

In this activity you will be given a set of directions several different times from your partners. A brown bag of contains 12 popsicle sticks, each marked with a distance (blocks) and a direction (N, S, E, & W) with an arrow. Place two coins of masking tape about 5 cm long on your table so that they form a"+". Where they cross will be your point of origin (the hotel).

Procedure
Have your partner reach into the bag and give you a stick at random. Place the stick with the tail of the arrow on the hotel (or origin) with the tip of the arrow pointing along the specified direction. Have your partner give you another stick. Place the tail of the second stick so that it touches the tip of the first stick. Pivot the second stick so that it points along its specified direction. Repeat until all the sticks are gone.

Place a piece of masking tape to mark the place where you end up. Measure the distance and direction from the hotel to where you end up. The directed distance from the hotel to the *Green House* is called the *displacement*.

    displacement = _____

Pick up all the sticks and put them back into the bag. This time you pull the sticks out of the bag and have your partner place them tail-to-tip just as you did. Measure the displacement.

Summing Up

1. How do your displacements compare?

_____

_____

2. How does the order the sticks are combined affect the displacement?

_____

_____

Vector Walk

CONCEPTUAL **Physics** | **Experiment**
Exponential Growth & Doubling Time

The Forgotten Fundamental

Purpose
To demonstrate the consequences of fixed growth of expending a finite resource.

Required Equipment and Supplies
calculator
wall clock, digital preferred (seconds only is ideal)
gallon milk jug of water
2 eye droppers
2 100-mL graduated cylinders
2 500 mL beakers
1 liter beakers
assorted containers
1 bucket (gallon)
5 gallon plastic trash container or larger

Discussion
This scientific activity relates directly to one of the most pressing problems facing humanity — a dilemma that affects every person on this planet — and yet only a few comprehend. Most people have no real understanding of it at all! You will be the exception, and *you* can make a difference. The problem is *exponential growth* and the population explosion, and its accompanying depletion of non-renewable resources such as oil, uranium, precious metals, etc. (See Appendix IV, *Conceptual Physics*.)

Doubling time is the time required for a quantity that is increasing in an exponential fashion to double. There is 100% more of the quantity after one doubling time as there was when that time interval began. The formula for the doubling time in years is:

$$\text{Doubling Time} = T = \frac{70}{\text{percent growth rate}}$$

The world population is increasing at an average rate of 1.7% per year — a rather benign figure — or so it would seem. But by using the formula, we divide 70 by 1.7 % per year to get 40 years which means that the world's population will double in just 40 years — to *twice* that of what it is today. Because comparable growth rates have been occurring for centuries, humankind is faced with a vexing dilemma. More people are alive today than have ever lived in all time before the beginning of the most recent doubling, which occurred within the past 40 years. The minimum resources required to support such masses of people are equivalent to those used by all the people who ever lived before 1950. The implication is, of course, that if this growth rate continues during the next 40 years we must find resources equivalent to all those used in all time up to the present. However, we have difficulty adequately providing for the world's population today — let alone one *twice* its present size. This is because no matter what the population becomes, there is only a *finite* amount of any non-renewable resource on this or any other planet.

When the death rate equals the birth rate, the population growth rate is zero. It is very likely that something will happen—starvation, deprivation, diseases, war, etc. so that these two rates eventually do equal each other. Zero population growth can be achieved by either *increasing* the death rate or *lowering* the birth rate. The choice is ours.

The Forgotten Fundamental

Procedure

Step 1: For this activity, you will need three other partners. The object is to almost fill a 5 gallon plastic trash container to the full line. You will add water every 10 seconds to the container. Start with one drop, doubling to two drops the next 10 seconds, doubling to four drops the next 10 seconds, and so on every subsequent 10 seconds, until the container is full.

1. What is the doubling time?

_____

Although it seems easy at first, filling the container in an orderly manner is going to require a combination of careful planning, technique, and team work to keep things going on schedule. Jot down your computations as well as your plan of attack. You will then make a plot of *Number of Drops vs. Doubling Times*.

Determine the number of drops in 1 mL (1 mL = 1 cm^3) using an eyedropper and a graduated cylinder. Count the number of drops it takes to fill it to the 2 mL mark. The total number of drops divided by 2 is the number of drops in one mL of water.

drops of water in one mL = _____

Step 2: Begin the experiment with one drop, then two, four, etc. You will soon have to switch to mL instead of drops, then liters instead of mL, and so on. You and your partners must devise a team strategy so each person on the team knows exactly when and what quantities of water to add to the container until it's full.

Although it might seem like it will take forever to fill the container, you'll fill it soon enough — in somewhat under 3 minutes!

2. How many liters does it take to fill the bucket to the full-line? How many drops?

_____

_____

Step 3: Fill in Table A until the total drops collected equals the capacity of the container.

Table A

DOUBLING TIMES	1	2	3	4	5	6	7	8	9	10	11	12	13	14	15	16	17	18
DROPS ADDED	2^1 2	2^2 4	2^3 8	2^4 16	2^5 32	2^6 64	2^7 128	2^8										
TOTAL DROPS COLLECTED	2^2-1 3	2^3-1 7	2^4-1 15	2^5-1 31														

Step 4: Make a graph from your Table. If a graphing program like *Data Plotter* is available, input your data on the computer. Plot the *Number of Drops* along the vertical axis and *Doubling Times* (seconds) along the horizontal axis. The vertical axis should be at least 15-20 cm long. Draw a smooth curve to fit the data points. Describe your graph.

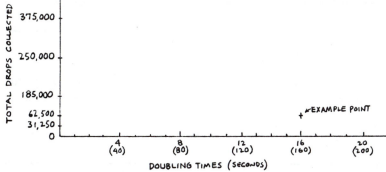

Step 5: Make another graph of the data using different coordinates. The number of drops in the container is 2^n, where n is the number of doubling times. If you are using *Data Plotter,* input the exponent of the power of 2 as the y-values for each data point (i.e., 1, 2, 3, ...etc.) What does your graph look like?

_____

_____

Summing Up

3. Discuss with your partners the implications this lab has on our society. Are governmental leaders aware of this problem? What parts of our government show strong similarities to what took place in this activity?

_____

_____

_____

_____

4. What do you think is the most humane way of reaching zero population growth? What steps do you think should be taken now? Later? Does waiting help solve the problem?

_____

_____

_____

_____

The Forgotten Fundamental

Going Further

5. You were probably surprised at how fast the container filled. When was most of the water added to the container?

_____

_____

Let's extend this activity by doing some computations.

6. Starting with a single drop and a doubling time of 10 seconds, how long would it take to fill a volume equal to a building 50 m x 20 m x 10 m with water?

_____

_____

7. The volume of the earth is about 10^{27} mL. How many doubling times would it take to fill a volume equal to the world?

_____

_____

8. When there were only 5,000 cases of AIDS in the USA, the disease spread throughout the population with a doubling time of about 6 months. How many doubling times would occur in 5 years?

_____

_____

Although unrealistic, assume no other factors impeded the spread of aids during that time, how many cases would occur in 5 years? 10 years?

_____

_____

9. Use *Data Plotter* to plot the United States postage rates versus the time they were in effect (Postage Rates vs. Time). How does your graph compare to those you made in Steps 4 and 5?

Date	Ra
July 1932	3
Aug 1958	4
Jan 1963	5
Jan 1968	6
May 1971	8
Mar 1974	10
Dec 1975	13
May 1978	15
Mar 1981	18
Nov 1981	20
Feb 1985	22
Apr 1988	25
Feb 1991	29

_____

_____

_____

Significant Figures and Uncertainty in Measurement

Units of Measurement

All measurements consist of a unit that tells what was measured and a number that tells how many units were measured. Both are necessary. If you say that a friend is going to give you 5, you are telling only *how many*. You also need to tell *what*: five fingers, five cents, five dollars, or five corny jokes. If your instructor asks you to measure the length of a piece of wood, saying that the answer is 26 is not correct. She or he needs to know whether the length is 26 centimeters, feet, or meters. All measurements must be expressed using a number and an appropriate unit. Units of measurement are more fully covered in your text.

Numbers

Two kinds of numbers are used in science — those that are counted or defined, and those that are measured. There is a great difference between a counted or defined number and a measured number. The exact value of a counted or defined number can be stated, but the exact value of a measured number cannot.

For example, you can count the number of chairs in your classroom, the number of fingers on your hand, or the number of quarters in your pocket with absolute certainty. Counted numbers are not subject to error.

Defined numbers are about exact relations and are defined to be true. The exact number of centimeters in a meter, the exact number of seconds in an hour, and the exact number of sides on a square are examples. Defined numbers also are not subject to error.

Every measured number, no matter how carefully measured, has some degree of uncertainty. What is the width of your desk? Is it 89.5 centimeters, 89.52 centimeters, 89.520 centimeters or 89.5201 centimeters? You cannot state its exact measurement with absolute certainty.

Uncertainty in Measurement

The uncertainty (or margin of error) in a measurement depends on the precision of the measuring device and the skill of the person who uses it. Uncertainty and error in laboratory measurements have a different meaning from "human error." The mistakes your lab partner may make in sloppy lab procedures are entirely different from the margin of error inherent in every measurement. Try as you may, you cannot entirely eliminate uncertainties that relate to the precision of your measuring instruments.

Fig. A

Uncertainty in a measurement can be illustrated by the two different metersticks in Figure A. The measurements are of the length of a table top. Assuming that the zero end of the meterstick has been carefully and accurately positioned at the left end of the table, how long is the table?

The upper scale in the figure is marked off in centimeter intervals. Using this scale you can say with certainty that the length is between 51 and 52 centimeters. You can say further that it is closer to 52 centimeters than to 53 centimeters; you can estimate it to be 51.2 centimeters.

The lower scale has more subdivisions and has a greater precision because it is marked off in millimeters. With this meterstick you can say that the length is definitely between 51.2 and 52.3 centimeters, and you can estimate it to be 52.25 centimeters.

Note how both readings contain some digits that are exactly known, and one digit (the last one) that is estimated. Note also that the uncertainty in the reading of the lower meterstick is less than that of the top meterstick. The lower meterstick can give a reading to the hundredths place, and top meterstick to the tenths place. The lower meterstick is more *precise* than the top one.

No measurements are exact. Measurements convey two kinds of information (1) the magnitude of the measurement and (2) the precision of the measurement. The location of the decimal point and

Significant Figures

Significant figures are the digits in any measurement that are known with certainty plus one digit that is estimated and hence is uncertain. The measurement 51.2 centimeters (made with the top meterstick in Figure A) has three significant figures, and the measurement 51.25 centimeters (made with the lower meterstick) has four significant figures. The right-most digit is always an estimated digit. Only one estimated digit is ever recorded as part of a measurement. It would be incorrect to report that in Figure A the length of the table as measured with the lower meterstick is 51.253 centimeters. This five-significant-figure value would have two estimated digits (the 5 and 3) and would be incorrect because it indicates a *precision* greater that the meterstick can obtain.

Standard rules have been developed for writing and using signifcant figures, both in measurements and in values calculated from measurements.

Rule 1
In numbers that do not contain zeros, all the digits are significant.

EXAMPLES:

4.1327 five significant figures
5.14   three significant figures
369   three significant figures

Rule 2
All zeros between significant digits are significant.

EXAMPLES:

8.052   four significant figures
7059   four significant figures
306   three significant figures

GOOD

POOR

Rule 3
Zeros to the left of the first nonzero digit serve only to fix the position of the decimal point and are not significant.

EXAMPLES:

0.0068   two significant figures
0.0427   three significant figures
0.000003   one significant figure

POO

POO

Rule 4
In a number with digits to the right of the decimal point, zeros to the right of the last non-zero digit are significant.

EXAMPLES:

53   two significant figures
53.0   three significant figures
53.00   four significant figures
0.00200   three significant figures
0.70050   five significant figures

Rule 5
In a number that has no decimal point and that ends in one or more zeros (such as 3600), the zeros that end the number may or may not be significant.

The number is ambiguous in terms of significant figures. Before the number of significant figures can be specified, further information is needed about how the number was obtained. If it is a measured number, the zeros are not significant. If the number is a defined or counter number, all the digits are significant.

Confusion is avoided when numbers are expressed in scientific notation. All digits are taken to be significant when expressed this way.

EXAMPLES:

4.6×10^{-5}	two significant figures
4.60×10^{-5}	three significant figures
4.600×10^{-5}	four significant figures
2×10^{-5}	one significant figure
3.0×10^{-5}	two significant figures
4.00×10^{-5}	three significant figures

Rounding Off

Calculators often times display eight or more digits. How do you round off such a display of digits to, say, three significant figures? Three simple rules govern the process of deleting unwanted (nonsignificant) digits from a calculator number.

Rule 1
If the first digit is to the right of the significant figure is less than 5, that digit and all the digits that follow it are simply dropped.

EXAMPLE:

51.234 rounded off to three signifcant figures becomes 51.2.

Rule 2
f the first digit to be dropped is a digit greater than 5, or if it is a 5 followed by digits other than zero, the excess digits are dropped and the last retained digit is increased in value by one unit.

EXAMPLE: 51.35, 51.359, and 51.3598 rounded off to three significant figures all become 51.4.

Rule 3
If the first digit to be dropped is a 5 not followed by any other digit, or if it is a 5 followed only be zeros, an odd-even rule is applied.

That is, if the last retained digit is even, its value is not changed, and the 5 and any zeros that follow are dropped. But if the last digit is odd, its value is increased by one. The intention of this odd-even rule is to average the effects of rounding off.

EXAMPLES:

74.2500 to three significant figures becomes 74.2.

89.3500 to three significant figures becomes 89.4.

Appendix

Significant Figures and Calculated Quantities

Suppose that you measure the mass of a small wooden block to be 2 grams on a balance, and you find that its volume is 3 cubic centimeters by poking it beneath the surface of water in a graduated cylinder. The density of the piece of wood is its mass divided by its volume. If you divide 2 by 3 on your calculator, the reading on the display is 0.7777777. However, it would be incorrect to report that the density of the block of wood is 0.7777777 gram per cubic centimeter. To do so would be claiming a degree of precision than is warranted. Your answer should be rounded off to a sensible number of significant figures.

The number of significant figures allowable in a calculated result depends on the number of significant figures in the data used to obtain the result, and on the type of mathematical operation(s) used to obtain the result. There are separate rules for multiplication and division, and for addition and subtraction.

Multiplication and Division

For multiplication and division an answer should have the number of significant figures found in the number with the fewest significant figures. For the density example, the answer would be rounded off to one significant figure, 0.7 gram per cubic centimeter. If the mass were measured to be 2.0 grams, and if the volume were still take to be 3 cubic centimeters, then the answer would still be rounded of to 0.7 gram per cubic centimeter. If the mass were measured to be 2.0 and the volume 3.0 or 3.00 cubic centimeters, the answer would be rounded off to two significant figures: 0.67 gram per cubic centimeter.

Study the following examples. Assume that the numbers being multiplied or divided are measured numbers.

EXAMPLE A:

8.536 x 0.47 = 4.01192 (calculator answer)

The input with the fewest significant figures is 0.47, which has two significant figures. Therefore, the calculator answer 4.01192 must be rounded off to 4.0.

EXAMPLE B:

3840 x 285.3 = 13.45916 (calculator answer)

The input with the fewest significant figures is 3840, which has three significant figures. Therefore, the calculator answer 13.45916 must be rounded off to 13.5.

EXAMPLE C:

360.0 ÷ 3.000 = 12 (calculator answer)

Both inputs contain four significant figures. Therefore, the correct answer must also contain four significant figures, and the calculator nswer 12 must be written as 12.00. In this case the calculator gave too few significant figures.

Addition and Subtraction

For addition or subtraction the answer should not have digits beyond the last digit position common to all the numbers being added and subtracted. Study the following examples:

EXAMPLE A:

34.6
18.8
<u>15</u>
67.4 (calculator answer)

The last digit position common to all numbers is the units place. Therefore, the calculator answer of 67.4 must be rounded off to the units place to become 67.

EXAMPLE B:

    20.02
    20.002
    <u>20.0002</u>
    60.0222 (calculator answer)

EXAMPLE C:

    345.56 - 245.5 = 100.06 (calculator answer)

The last digit position common to both numbers in this subtraction is the tenths place. Therefore, the answer should be round off to 100.1.

Percentage Error

If your aunt told you that she had made $500 in the stock market, you would be more impressed if this gain were on a $500 investment than if it were on a $5,000 investment. In the first case she would have doubled her investment and made a 100% gain. In the second case should would have made only a 1% gain.

In laboratory measurements it is the *percentage difference* that is important, *not the size of the difference*. Measuring something to within 1 centimeter may be good or poor, depending on the length of the object you are measuring. Measuring the length of a 10-centimeter pencil to +/– one centimeter is quite a bit different from measuring the length of a 100-meter track to the same +/– one centimeter. The measurement of the pencil shows a relative uncertainty of 10%. The track measurement is uncertain by only 1 part in 5,000, or 0.01%.

The relative uncertainty or relative margin of error in measurements, when expressed as a percentage, is often called the *percentage of error*. It tells by what percentage a quantity differs from a known accepted value as determined by skilled observers using high precision equipment. It is a measure of the accuracy of the method of measurement as well as the skill of the person making the measurement. The percentage of error is found by dividing the difference between the measured value and the accepted value of a quantity by the accepted value, and then multiplying this quotient by 100%.

$$\% \text{ error} = \frac{(\text{accepted value} - \text{measured vlaue})}{(\text{accepted value})} \times 100\%$$

For example, suppose that the measured value of the acceleration of gravity is found to be 9.44 ms2. The accepted value is 9.81 m/s^2. The difference between these two values is (9.81 m/s^2) – (9.44 m/s^2), or 0.37 m/s^2.

$$\% \text{ error} = \frac{0.37 \text{ m/s}^2}{9.81 \text{ m/s}^2}$$

$$= 3.77 \%$$